DISEASES OF VEGETABLE CROPS IN AUSTRALIA

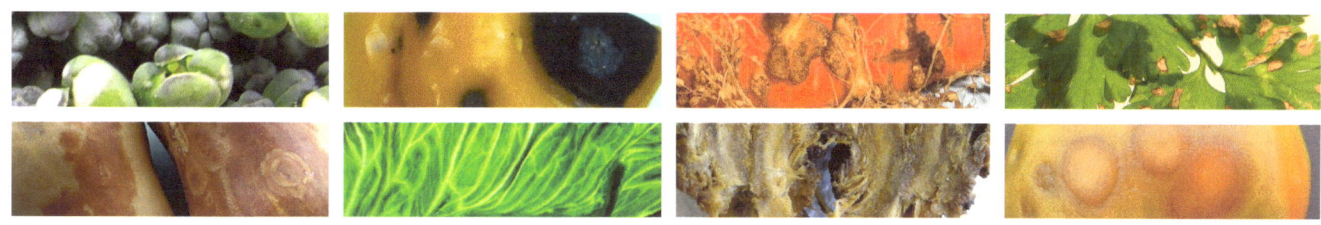

DENIS PERSLEY, TONY COOKE AND SUSAN HOUSE

Queensland Primary Industries and Fisheries,
Department of Employment, Economic Development and Innovation

CSIRO

PUBLISHING

Queensland Government

National Library of Australia Cataloguing-in-Publication entry

Diseases of vegetable crops in Australia / editors: Denis Persley, Tony Cooke, Susan House.

9780643096387 (hbk.)

Includes index.
Bibliography.

Vegetables – Diseases and pests – Australia.

Persley, Denis M.
Cooke, Tony
House, Susan.

635.0493

Published by

CSIRO PUBLISHING
36 Gardiner Road, Clayton VIC 3168
Private Bag 10, Clayton South VIC 3169
Australia

Telephone: [+613] 9545 8555
Local call: 1300 788 000 (Australia only)
Fax: +61 3 9662 7555
Email: csiropublishing@csiro.au
Web site: www.publishing.csiro.au

Front cover (clockwise from top left): White blister on broccoli (Leif Forsberg/QPIF); Anthracnose on capsicum (Tony Cooke/ QPIF); Nematode galling on carrot (Jenny Cobon/QPIF); Early blight on celery (Leif Forsberg/QPIF); Tomato spotted wilt on ripe tomato (Murray Sharman/QPIF); Root rot on parsley (Leif Forsberg/QPIF); Big-vein disease on lettuce (Cherie Gambley/QPIF); Brown etch on pumpkin (Leif Forsberg/QPIF). Background image by Ted Hamilton.

Set in 10.5/12.5 Minion
Edited by Janet Walker
Cover design by The Modern Art Production Group
Text design by James Kelly
Typeset by Planman Technologies India Pvt. Ltd.
Index by Indexicana
Printed by Ingram Lightning Source

CSIRO PUBLISHING publishes and distributes scientific, technical and health science books and journals from Australia to a worldwide audience and conducts these activities autonomously from the research activities of the Commonwealth Scientific and Industrial Research Organisation (CSIRO).

Feb26_RP_ILS

CONTENTS

FOREWORD

Australia is largely self-sufficient in fruit and vegetables. The industry is intensive, typically seasonal in operation, and dominated by small-scale farms. With fruit, nut and vegetable cropping worth $5.5 billion, and other horticulture an additional $1.7 billion in 2005–06, horticulture is Australia's third-largest agricultural industry, and a major employer in rural areas.

Most Australian production is for domestic markets, but the horticulture industries also contribute significantly to exports. In 2004–05, fresh fruit, vegetable and nut exports were worth $800 million, wine exports were worth $2.7 billion, and other processed horticultural produce was worth $400 million. At the same time, fruit, nut and vegetable imports provided additional diversity and met gaps in domestic production.

Effective, integrated disease-management and quarantine strategies have been essential for minimising losses and facilitating market access for Australia's development as a major producer and exporter of high-quality fruit, nuts and vegetables. In recent years, the threats of global warming to the sustainability of cropping, the limitations in water and land availability, and the rising costs of fuel and other inputs, have reinforced the need for making production and marketing as efficient as possible through better disease control. Also, in line with community expectations for high-quality produce with minimal chemical residues, it has become critical for the rural community to manage pests and diseases responsibly, as well as effectively.

Accurate identification of the plant diseases attacking horticultural crops is a key step in choosing safe and effective control options and in maintaining biosecurity preparedness. This need was met by publishing the *Handbook of Plant Diseases in Colour – Fruit and Vegetables* by the Queensland Department of Primary Industries in 1978, and in subsequent separate volumes covering fruit and vegetables respectively in 1993 and 1994. This current, complete revision, *Diseases of Vegetable Crops in Australia,* incorporates overviews of the causes and the main disease types attacking vegetable crops, and 21 separate chapters covering particular vegetable groups. The book illustrates the major diseases of vegetables currently present in Australia, with concise information on their cause, symptoms, disease cycle and management. In addition, some key exotic diseases that represent biosecurity threats to Australian horticulture are covered.

Assembling the illustrations and preparing the text has been a national effort. Although most of the authors are from Queensland Primary Industries and Fisheries, plant pathologists in all States have contributed either as authors or by providing images and information. The Australasian Plant Pathology Society, which represents the discipline of plant pathology in Australia and New Zealand, is very pleased to provide sponsorship and support for this important publication.

Previous editions in this series have proved popular and I am sure that this book will prove an essential resource for anyone involved in vegetable production and quality management.

Dr. Greg I Johnson
President
Australasian Plant Pathology Society

Plant Health is Earth's Wealth

PREFACE

Diseases of Vegetable Crops in Australia is the third in a series of plant disease handbooks produced by plant pathologists from Queensland Primary Industries and Fisheries. The purpose of each has been to provide a diagnostic guide and a key reference for diseases affecting vegetable crops in Australia. The first, *Handbook of Plant Diseases in Colour – Fruit and Vegetables,* was published by the Queensland Department of Primary Industries in 1978 with a second edition published in 1982. It was fully revised in the mid 1990s and published as two separate volumes, *Diseases of Fruit Crops* and *Diseases of Vegetable Crops.*

This current edition, written some 15 years later, is extensively revised and expanded.

With the collaboration of colleagues throughout Australia, the editors and authors provide essential information about the important diseases affecting most vegetable crops grown across Australia's diverse production areas. The disease descriptions are supported by many diagnostic images.

Diseases of Vegetable Crops in Australia is a valuable guide for growers, their consultants and managers, horticulturists, plant pathologists, plant protection diagnosticians, integrated pest-management specialists, educators, students and agribusiness representatives, as well as the enthusiastic home gardener and hobby farmer.

The first chapter is an introduction to the causes, nature and principles underlying the control of plant diseases and includes other sources of information about pathogen groups and plant diseases. The second chapter is an overview of several diseases affecting many vegetable crops and links to subsequent chapters on specific crop diseases.

The remaining chapters discuss the diseases of specific crops. The diseases are organised first by pathogen type then by an alphabetical listing of diseases under each type of pathogen.

Each disease description includes information about symptoms, means of spread, disease development and survival, importance and management. The latter emphasises the need to adopt an integrated approach to disease management, applying all the appropriate cultural, crop-management and chemical methods to achieve a cost-effective and sustainable result.

Sources of further information are given at the end of each chapter.

The colour images help with disease identification, and include early symptoms or distinguishing features.

Early detection of exotic diseases that are a biosecurity threat to Australian horticultural industries is vital if they are to be contained or appropriately managed. To assist industry awareness, the major biosecurity threats for most crops have been included.

A glossary and index complete the book.

Specific chemical recommendations are not included in the disease-management sections because they change regularly, and can vary between regions and States. Current chemical recommendations can be found in specific crop management guides, and by government and private extension and consultancy services.

CONTRIBUTORS

Chrys Akem, Queensland Primary Industries and Fisheries, Department of Employment, Economic Development and Innovation, Queensland.
Email: chrys.akem@deedi.qld.gov.au

Desmond Auer, Department of Primary Industries, Victoria.
Email: desmond.auer@dpi.vic.gov.au

Eric Coleman, Queensland Primary Industries and Fisheries, Department of Employment, Economic Development and Innovation, Queensland.
Email: eric.coleman@deedi.qld.gov.au

Barry Conde, Northern Territory Government, Department of Regional Development, Primary Industry, Fisheries and Resources.
Email: Barry.Conde@nt.gov.au

Tony Cooke, Queensland Primary Industries and Fisheries, Department of Employment, Economic Development and Innovation, Queensland.
Email: tony.cooke@deedi.qld.gov.au

Bob Davis, Department of Primary Industries and Fisheries, Queensland (now retired).

Rudolf deBoer, Department of Primary Industries, Victoria.
Email: Dolf.deBoer@dpi.vic.gov.au

Leanne Forsyth, NSW Department of Industry and Investment.
Email: leanne.forsyth@industry.nsw.gov.au

Cherie Gambley, Queensland Primary Industries and Fisheries, Department of Employment, Economic Development and Innovation, Queensland.
Email: cherie.gambley@deedi.qld.gov.au

Barbara Hall, South Australian Research and Development Institute, Primary Industries and Resources, South Australia.
Email: Barbara.Hall@sa.gov.au

Christine Horlock, Queensland Primary Industries and Fisheries, Department of Employment, Economic Development and Innovation, Queensland.
Email: christine.horlock@deedi.qld.gov.au

Heidi Martin, formerly Department of Primary Industries and Fisheries, Queensland.

Elizabeth Minchinton, Department of Primary Industries, Victoria.
Email: liz.minchinton@dpi.vic.gov.au

Ken Pegg, Queensland Primary Industries and Fisheries, Department of Employment, Economic Development and Innovation, Queensland.
Email: ken.pegg@deedi.qld.gov.au

Denis Persley, Queensland Primary Industries and Fisheries, Department of Employment, Economic Development and Innovation, Queensland.
Email: denis.persley@deedi.qld.gov.au

Murray Sharman, Queensland Primary Industries and Fisheries, Department of Employment, Economic Development and Innovation, Queensland.
Email: murray.sharman@deedi.qld.gov.au

Graham Stirling, Biological Crop Protection Pty Ltd, Moggill, Queensland.
Email: graham.stirling@biolcrop.com.au

Len Tesoriero, NSW Department of Industry and Investment.
Email: len.tesoriero@industry.nsw.gov.au

John Thomas, Queensland Primary Industries and Fisheries, Department of Employment, Economic Development and Innovation, Queensland.
Email: john.thomas@deedi.qld.gov.au

ACKNOWLEDGEMENTS

We wish to acknowledge funding support from Horticulture Australia Limited (HAL) and the Australasian Plant Pathology Society (APPS).

Know-how for Horticulture™

We thank Dr Ian Porter and Leanne Wilson for their support in initiating the project to allow this revision.

The editors and authors are grateful for the assistance and advice provided by colleagues throughout Australia, especially Andre Drenth, John Duff, Leif Forsberg, Leanne Forsyth, Cherie Gambley, Des McGrath, Ken Pegg, Roger Shivas, Len Tesoriero, John Thomas, Cherie Thomson, Scott Weir, Andrew Watson and Carrie Wright.

Illustrations were supplied by Paul Mooney of Mooney Fine Arts. Special thanks to Leif Forsberg for excellent photographic work.

Photographic images were provided by the following people and organisations:

Queensland Primary Industries and Fisheries (QPIF), Department of Employment, Economic Development and Innovation:

Denis Persley, Heidi Martin and Tony Cooke – principal contributors

Jay Anderson – Fig. 18.11

Dean Beasley – Figs. 1.10, 4.12, 4.13

Jenny Cobon – Fig. 2.36

Eric Coleman – Figs. 22.1, 22.8

Peter Deuter – Fig. 1.1 (left)

Andre Drenth – (QPIF and Tree Pathology Centre, University of Queensland) Figs. 2.19, 18.17, 19.18, 19.19, 23.33

Cherie Gambley – Figs. 1.1 (centre), 2.5, 2.15, 11.67, 15.22

Leif Forsberg – Figs. 1.12, 1.5, 1.9, 1.13, 2.1, 2.4, 5.19, 7.28, 11.25, 12.7, 14.25, 23.8

Jerry Lovatt – Figs. 22.2, 22.3, 22.4

Elio Jovicich – Fig. 11.44

Gerry Macmanus – Figs. 12.1, 16.28

Russel McCrystal – Fig. 22.6

Des McGrath – Fig. 23.10

Alistair McTaggart – Fig. 1.9 (inset)

Murray Sharman – Figs. 1.2 (centre), 1.4 (right), 23.65

Mike Smith – Figs. 13.7, 13.8

John Thomas – Fig. 1.17

Ian Walker – Figs. 1.14, 2.20, 2.46, 8.1, 23.19, 23.20, 23.27, 23.38, 23.39, 23.40

Anthony Young – Figs. 19.3, 19.4, 23.1, 23.15

Other organisations:

Arie Baelde, RijkZwaan Australia – Figs. 14.35, 23.61

Afshen Shamshad, University of Sydney – Fig. 14.17

Robin Coles, Rural Solutions SA, Primary Industries South Australia – Fig. 9.4

Barry Conde, Northern Territory Department of Primary Industries – Figs. 3.18, 3.19

Brenda Coutts, Department of Agriculture and Food, Western Australia – Figs. 8.22, 9.5, 9.14, 9.15, 19.45, 23.54, 23.55

Dolf de Boer, Department of Primary Industries, Victoria – Figs. 19.10, 19.13, 19.20, 19.24, 19.29, 19.30

Kai-Shu Ling, US Department of Agriculture, Agricultural Research Service, S. Carolina, USA – Fig. 23.62

Solke H DeBoer, Centre for Animal and Plant Health, Charlottetown, PE, Canada – Figs. 1.19 (left), 19.1, 19.2

Caroline Donald, Department of Primary Industries, Victoria – Figs. 7.10, 7.10 (detail), 7.14

John Fletcher, New Zealand Institute for Plant and Food Research Limited – Figs. 19.22, 19.42

Leanne Forsyth, NSW Department of Industry and Investment – Figs. 15.28, 7.31

Kaye Ferguson, South Australian Research and Development Institute, PISA – Figs. 15.6–15.7

Helen Grogan Teagasc, Kinsealy Research Centre, Dublin, Ireland – Figs. 14.15, 14.16, 14.18, 14.19

Barbara Hall, South Australian Research and Development Institute, PISA – Figs. 16.1–16.3, 16.15, 16.16, 16.20, 16.22, 16.25

Ian Porter, Department of Primary Industries, Victoria – Figs. 3.9, 7.3, 7.15, 7.20–7.24

Graham Jackson – Figs. 3.20, 3.21

Jose Liberato, Northern Territory Department of Primary Industries – Fig. 5.27

Margaret McGrath, Department of Plant Pathology and Plant Microbe Biology, Ithaca, NY, USA – Figs. 1.19 (crop), 8.10–8.12, 11.15–11.17, 11.46–11.49

Sue Pederick, South Australian Research and Development Institute – Fig. 23.42

Liz Minchinton & Desmond Auer, Department of Primary Industries, Victoria – Figs. 17.1–17.6

Howard Schwartz, Colorado State University, Fort Collins, USA – Figs. 16.4, 16.5, 16.7, 16.8

Afshen Shamshad – Fig. 14.17

Graham Stirling, Biological Crop Protection, Queensland – Figs. 2.39, 13.9, 13.10

The American Phytopathological Society (APS) – Figs. 15.14 (JC Hubbard), 19.41 (H. Torries)

Len Tesoriero and Lowan Turton, NSW Department of Primary Industries – Figs. 3.1–3.3, 3.5–3.7, 3.10, 3.14, 3.17

Dan Trimboli – Fig. 15.26

Oscar Villalta, Department of Primary Industries, Victoria – Fig. 2.31

Lisa Ward, MAF Biosecurity, New Zealand – Figs. 2.6–2.9

Andrew Watson, NSW Department of Primary Industries – Figs. 2.18, 5.31, 9.8, 9.9, 18.06, 18.07, 18.08, 21.6

Calum Wilson, Tasmanian Institute of Agricultural Research, Tasmania – Figs. 19.42–19.44

Humankind has struggled with plant diseases since the dawn of agriculture. There are references in the Old Testament to the ravages of rust and blight on cereals and grapevines in the Ancient world. The potato famine in Ireland in the 1840s led to the mass migration of Irish refugees to Australia, Britain and North America so that today, almost one in 10 Australians can trace their ancestry back to Ireland. The cause of the famine was the potato blight pathogen *Phytophthora infestans*, which destroyed plants and tubers under prolonged wet and cold weather.

Plant diseases are closely connected with current issues facing agriculture and the environment. Global warming and rainfall reliability will have a considerable influence on disease distribution and severity in crops. Plant diseases are a major factor in world food security, and biosecurity issues are a key component in international trade agreements. The absence of many damaging pathogens in Australia provides competitive advantage in trade and access to new markets. Furthermore, savings in costs associated with managing or eradicating these pathogens means that, in some cases, production is more efficient and sustainable than in overseas countries.

■ CAUSES OF DISEASE IN PLANTS

A simple definition of a plant disease is any disturbance that interferes with the plant's normal structure, function or economic value. Plant diseases divide conveniently into (a) those caused by parasitic microorganisms or pathogens, and (b) non-parasitic diseases or disorders. These latter include mineral excesses and imbalances, incorrect storage conditions after harvest, environmental influences (such as atmospheric pollutants) and herbicide damage. Table 1.1 lists some physiological disorders of vegetable crops.

This handbook concerns diseases caused by pathogens. The major groups of plant pathogens are fungi, bacteria, viruses

Fig 1.1 Healthy produce is a team effort between growers, horticulturalists and plant pathologists.

Table 1.1 Some physiological disorders of vegetable crops

CROP	DISORDER	CAUSE
Lettuce	Tipburn	Associated with a low concentration of calcium in the tissues Promoted by conditions favouring rapid plant growth: warm temperatures, excess fertilisation and high light intensity
Cucurbits	Measles	Guttation droplets on the fruit surfaces Extended periods of high humidity, or wide temperature fluctuations between night and day
Cucurbits	Premature fruit abortion	Poor pollination
Capsicum	Yolo spot	Varietal susceptibility Calcium deficiency
Tomato	Blossom end rot	Localised calcium deficiency in the distal end of the fruit
Carrots	Black ring	A combination of varietal, physiological and fungal factors

Fig 1.2 Disease is a constant threat to profitable vegetable production. Left to right: downy mildew in cucurbits, *Tomato spotted wilt virus* and Sclerotinia drop in lettuce.

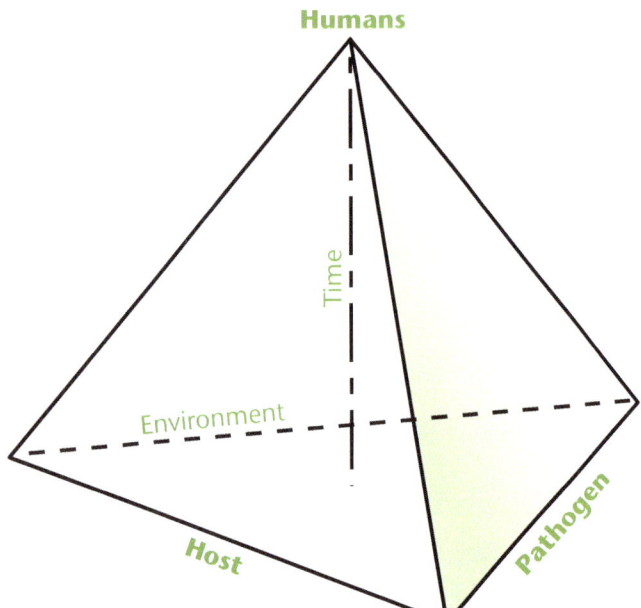

Fig 1.3 Interrelationships in plant disease epidemics.

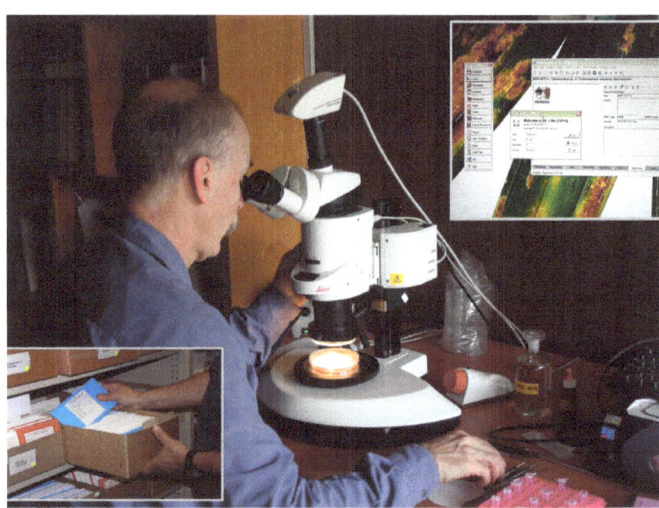

Fig 1.5 Modern herbaria maintain state-of-the art electronic databases for original specimens and use facelift packaging.

Fig 1.4 The aim of the Australian Vegetable Industry Biosecurity Plan is to reduce the risk of exotic disease introductions to our local crops. Threats include: bacterial ring rot (left) and whitefly-transmitted Geminiviruses, which are common throughout Asia (right).

Fig 1.6 Cooperative research is important for Australia's farming future. It allows us to develop a better understanding of possible threats to our vegetable industries.

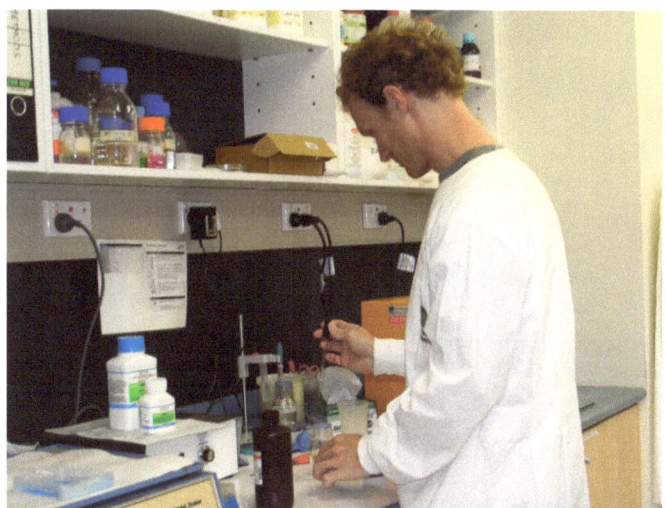

Fig 1.7 Conventional research and modern molecular techniques enable today's plant pathologists to develop management recommendations for plant pathogen outbreaks.

and nematodes. All diseases caused by pathogens are the result of an interaction between a susceptible plant, a pathogen capable of causing disease and a favourable environment (Fig 1.3).

The following sections outline key characteristics of plant pathogens.

■ FUNGI

Fungi are mostly filamentous organisms that lack the green pigment chlorophyll and must obtain energy from the material on which they grow. Most fungi are saprophytes, living entirely on dead or decaying organic matter. Fungi are the most important and common cause of plant disease, with about 23 000 species known to infect plants, although it has been estimated that the actual number may be as high as 270 000 species. Some fungal pathogens can survive only by growing in their living host plants; these are termed obligate parasites or biotrophs. Examples include the rusts, smuts and powdery mildews.

The majority of fungal pathogens are non-obligate or facultative parasites requiring a living host plant for only part of their life cycle. Fungi consist of individual living filaments called hyphae, which collectively form mycelium. As in other organisms, reproduction is an essential part of the life cycle of a fungus. Most fungi have the ability to reproduce both sexually or asexually. Usually, the asexual (imperfect, anamorphic) stage is the active pathogen and the sexual (perfect, teleomorphic) stage may occur only rarely. The sexual stage helps the fungus survive adverse, often seasonal conditions, and provides genetic diversity for the organism. The basic reproductive unit of fungi is

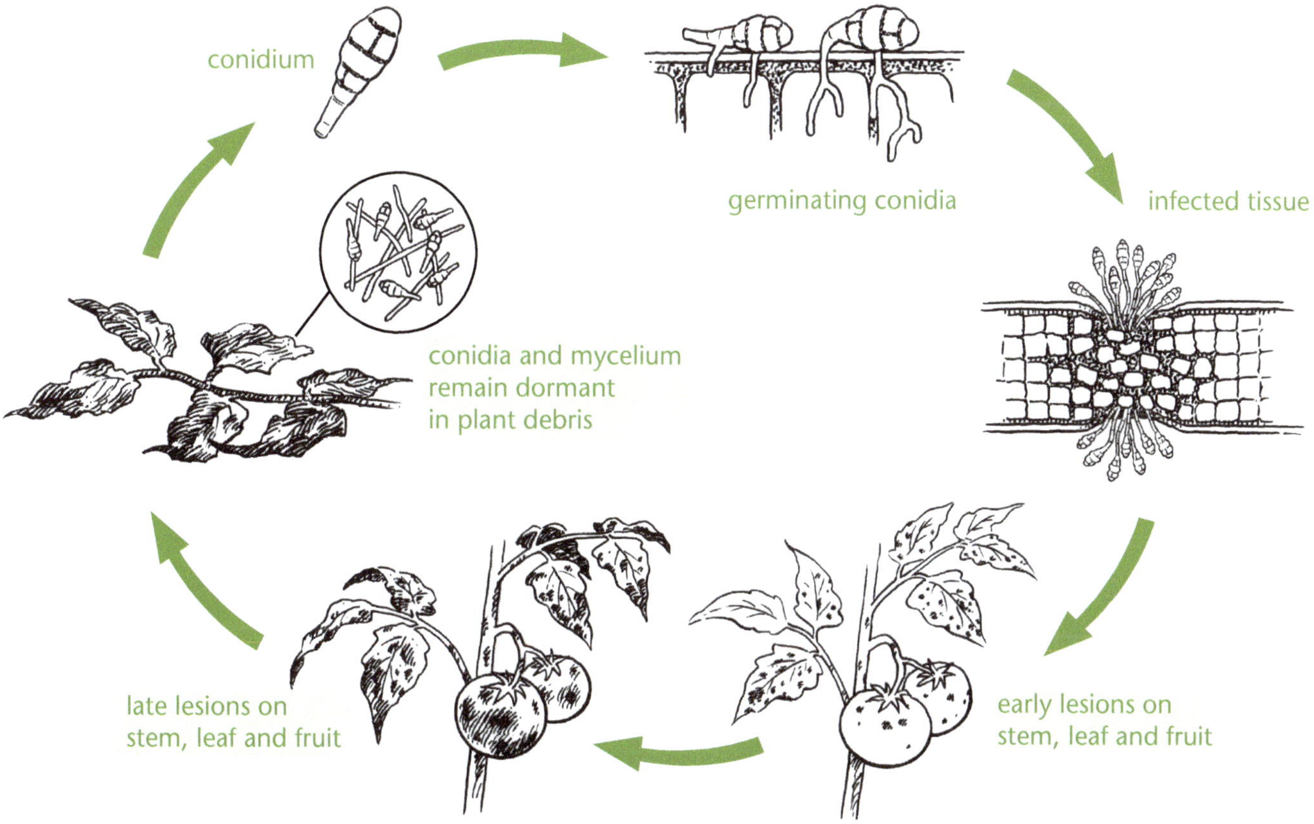

conidium

germinating conidia

infected tissue

conidia and mycelium remain dormant in plant debris

early lesions on stem, leaf and fruit

late lesions on stem, leaf and fruit

Fig 1.8 Disease cycle of *Alternaria* species.

Fig 1.9 Symptoms of Alternaria on tomato leaf and fruit and a magnified view of the spores.

Fig 1.10 Rust on asparagus fern. Inset: Rust uredinospores.

Fig 1.11 Powdery mildew on a pumpkin leaf. Inset: Spore-bearing structures.

the spore, which germinates to produce hyphae. Spores may result from both asexual and sexual reproduction and often a single fungal species may produce several different types of spores. The sexual stage of some fungi is unknown, or may not even exist, and so only the asexual stage is known. Formerly called *Fungi Imperfecti,* most are actually ascomycetes or basidiomycetes.

Spores develop in special structures called fruiting bodies, which provide some protection against desiccation and ultraviolet radiation. Often produced in enormous numbers, spores disperse by wind currents, rain, running water or insects. Thick-walled spores resistant to adverse conditions allow fungi to survive for long periods in the soil and on both living and dead plants.

Fungal mycelium may also form small, hard structures called sclerotia, which are important survival structures for many fungi.

The true fungi belong in their own kingdom, the Fungi. Now we know that the true fungi are more closely related

to animals than to plants. The four major groups (phyla) of true fungi are the Ascomycota, Basidiomycota, Chytridimycota and Glomeromycota (includes traditional Zygomycota). The first two groups contain most of the plant pathogenic fungi.

Some important and widespread plant pathogens known as oomycetes were once considered part of the fungal kingdom. Their name derives from the round oospores produced by sexual reproduction and serving as thick-walled survival spores. Oomycetes are adapted to living in moist environments and produce asexual zoospores with flagella that allow them to swim. The best-known oomycetes are the plant pathogens *Phytophthora*, *Pythium* and the downy mildews. Structural, molecular and biochemical studies show that oomycetes are more closely related to diatoms, kelps and golden-brown algae. These organisms are now placed into the Kingdom Stramenopila.

The main characteristics of the plant pathogenic fungi and oomycetes are listed in Table 1.2.

Fig 1.12 Fungi and bacteria are major causes of disease in vegetable crops. Most will grow in culture.

Table 1.2 Fungal and fungal-like pathogens

GROUP (PHYLA)		DISTINGUISHING FEATURES	REPRESENTATIVE MEMBERS
True fungi	Chytridiomycota (Chytridiomycetes)	Motile oospores; survive as resting sporangia; most species saprobic	*Physoderma, Synchytrium, Olpidium*
	Glomeromycota (includes traditional Zygomycota)	Sexual spores: zygospores; asexual spores; sporangiospores from sporangia	*Rhizopus, Mucor*
	Ascomycota (Ascomycetes)	Sexual spores; (ascospores) form in an ascus; asexual spores are conidia; septate hyphae	Powdery mildews, *Cercospora, Guignardia, Meliola, Taphrina, Septoria, Venturia*
	Basidiomycota (Basidiomycetes)	Sexual spores (basidiospores) produced in a basidium; clamp connections develop at hyphal septa	Rusts, smuts, mushrooms, *Armillaria*
Fungal-like organisms	Stramenopila (Stramenopiles) Oomycota (Oomycetes)	Non-septate hyphae cell wall contains cellulose and glucans instead of chitin as in true fungi; sexual spores are oospores; asexual spores are zoospores (two flagellae) produced in a sporangium	*Phytophthora, Pythium, Albugo,* downy mildews
	Protozoa Plasmodiophoromycota	Intracellular in algal, fungal or plant hosts; develops a multinucleate, unwalled plasmodium in host cell; flagellate zoospores; common in soil and aquatic habitats	*Plasmodiophora brassicae* — club root of brassicas

■ BACTERIA AND PHYTOPLASMAS

Bacteria are tiny, single-celled organisms that lack chlorophyll. Cells reproduce rapidly by dividing into two (fission). Many bacteria produce extracellular polysaccharides that form a slime layer or capsule around the cell, which assists in plant infection. Members of the genera *Agrobacterium, Clavibacter, Erwinia, Pseudomonas, Xanthomonas, Ralstonia* and *Xylella* account for most species that cause diseases in plants.

Most bacterial plant pathogens are facultative parasites that adapt readily to different environments and can usually be cultured easily in the laboratory. Some bacteria that infect plants have never been cultured, or only with great difficulty, using specialised media. These are known as fastidious bacteria. They include the xylem-limited *Xylella fastidiosa*, the cause of Pierce's disease of grapevine, and the phloem-limited *Spiroplasma citri*, the cause of citrus stubborn disease.

Bacteria can survive for some time on plant surfaces as epiphytes, becoming active when conditions favour their development. The organisms can also survive in soil and crop debris, and in seeds and other plant parts. With the exception of *Streptomyces*, plant bacterial pathogens do not form spores.

Bacteria spread in infected seed and propagating material, water splash and wind-driven rain. Overhead irrigation is often an important means of spreading bacteria within a crop. Bacteria also spread with insects and with workers and machinery moving through a crop that is wet from rain or dew. Some species such as *Xylella* have specific insect vectors.

Bacteria infect plants through wounds or natural openings such as stomata and hydathodes. Warm, wet weather favours their development, whereas growth is often arrested by hot, dry conditions.

Fig 1.13 Bacteria growing in culture showing bacterial ooze from plant tissues (left) and a cell viewed through the microscope (right).

Fig 1.14 A tomato plant infected with tomato big bud disease showing distorted, new growth and the phytoplasma pathogen in an infected cell.

Phytoplasmas, previously called mycoplasma-like organisms, are similar to the true bacteria. They are of various shapes, including spherical, ovoid and filamentous, and lack a rigid cell wall. Phytoplasmas are spread by sap-sucking leafhoppers and planthoppers and infect only the phloem tissue of plants.

Typical diseases caused by phytoplasmas are the 'big bud' and 'little leaf' diseases of many crop and weed plants.

■ VIRUSES AND VIROIDS

Viruses are minute, non-cellular, obligate parasites consisting of a nucleic acid core, which contains the genetic information necessary for replication, surrounded by a protective protein or lipoprotein coat. Viruses cannot reproduce outside a host cell and they use the plants cell structures and components to produce more virus particles, to the detriment of the plant.

Many plant viruses are spread by sap-sucking insects, in particular, aphids, leafhoppers, thrips and whiteflies. All have piercing-sucking mouthparts including needle-like stylets which allow the insects to feed on the contents of plant cells. Transmission is an intricate biological process, often requiring the virus to form a close relationship with insect tissues before transmission is possible. Particular viruses are, almost always, spread by only one insect type. For example, aphids can transmit *Papaya ringspot virus* but not *Tomato yellow leaf curl virus*, which, in turn, is transmitted only by the silverleaf whitefly.

Three broad categories of insect transmission of plant viruses are recognised: non-persistent, semi-persistent and

Fig 1.15 Plant viruses viewed through an electron microscope showing isometric, flexuous and bullet-shaped particles.

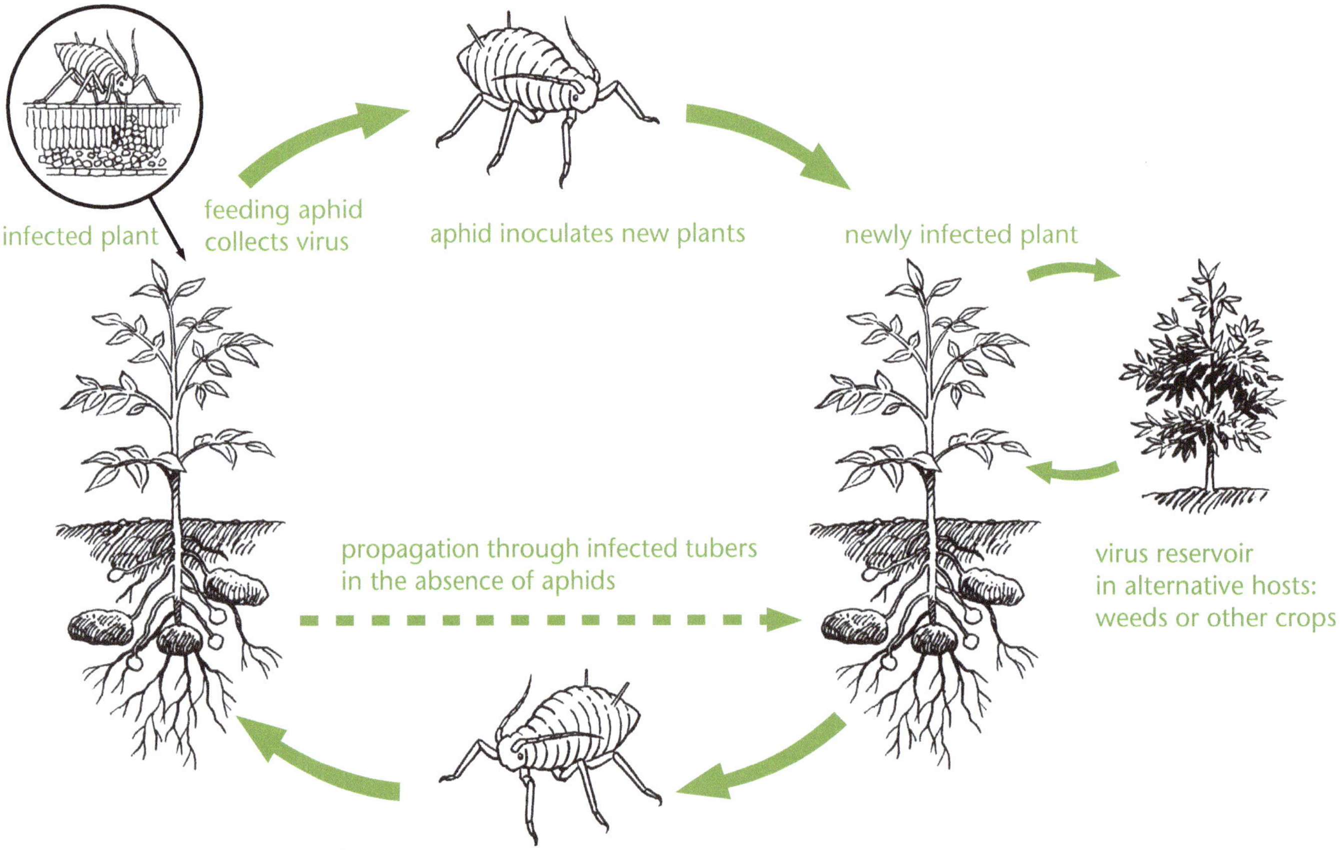

infected plant

feeding aphid collects virus

aphid inoculates new plants

newly infected plant

virus reservoir in alternative hosts: weeds or other crops

propagation through infected tubers in the absence of aphids

Fig 1.16 Disease cycle of *Potato leafroll virus*.

persistent or circulative transmission. The terms relate to the length of time an insect takes to acquire and transmit a virus, and the length of time the insect remains capable of transmitting the virus.

In non-persistent transmission the virus can be acquired from an infected plant or transmitted to another plant in less than one minute; the virus particles are usually retained on the insect's stylets for only a few hours. After this time, the insect obtains more virus only by feeding again on an infected plant.

In semi-persistent transmission, a virus can be acquired after 15 to 20 minutes of feeding, and the ability to transmit is retained for several days.

In persistent or circulative transmission, the insect needs to feed for up to several hours on an infected plant to acquire the virus, which then must circulate through the insect's body to the salivary glands for transmission to occur. This means there is a latent period, or time during which transmission cannot occur, while the virus particles travel through the

Fig 1.17 Field symptoms of *Potato leafroll virus*.

insect's body. When the latent period is completed the insect can transmit the virus for many weeks, or the rest of its life, without collecting more virus from an infected plant.

Table 1.3 Important viruses in Australian vegetable crops

VIRUS/VIRUS GROUP	MEANS OF SPREAD	IMPORTANT HOST PLANTS
Bean common mosaic virus (Potyvirus)	Seed, aphids (non-persistent)	Beans
Bean yellow mosaic virus (Potyvirus)	Aphids (non-persistent)	Legumes, some ornamentals
Beet western yellows virus (Polerovirus)	Aphids (persistent)	Brassicas, lettuce, legumes, brassica weed species
Capsicum chlorosis virus (Tospovirus)	Thrips (three species)	Capsicum, tomato, peanut
Carrot virus Y (Potyvirus)	Aphids (non-persistent)	Carrot
Celery mosaic virus (Potyvirus)	Aphids (non-persistent)	Celery
Cucumber mosaic virus (Cucumovirus)	Seed, vegetative propagation, aphids (non-persistent)	Wide host range including legumes, cucurbits, capsicum, tomato, lettuce, ornamentals, weeds
Iris yellow spot virus (Tospovirus)	Thrips (*Thrips tabaci*)	Onion
Johnson grass mosaic virus (Potyvirus)	Aphids (non-persistent)	Sweet corn, maize, sorghum
Lettuce mosaic virus (Potyvirus)	Lettuce seed, aphids (non-persistent)	Lettuce
Mirafiori lettuce virus (Ophiovirus)	Zoospores of the soil-borne fungus *Olpidium virulentus*	Lettuce
Papaya ringspot virus – Type W (Potyvirus)	Aphids (non-persistent)	Cucurbits
Pea seed-borne mosaic virus (Potyvirus)	Pea seed, aphids (non-persistent)	Pea and several other legumes
Potato leaf roll virus (Polerovirus)	Aphids (persistent), vegetative propagation (tubers)	Potato, tomato
Potato virus Y (Potyvirus)	Aphids (non-persistent)	Potato, tomato, capsicum
Squash mosaic virus (Comovirus)	Seed, several leaf chewing beetles	Cucurbits
Subterranean clover stunt virus (Nanovirus)	Aphids (persistent)	Legumes, including beans, pea, broad beans
Sweet potato feathery mottle virus (Potyvirus)	Vegetative propagation (cuttings, roots); aphids (non-persistent)	Sweetpotato
Tomato mosaic virus (Tobamovirus)	Seed, contact by handling, contaminated implements	Tomato
Tomato spotted wilt virus (Tospovirus)	Thrips (persistent, propagative)	Wide range of hosts among vegetable, ornamental and weed species
Tomato yellow leaf curl virus (Begomovirus)	Silverleaf whitefly (*Bemisia tabaci*) (persistent)	Tomato, bean, capsicum, several weed species
Turnip mosaic virus (Potyvirus)	Aphids (non-persistent)	Brassicas, lettuce, rhubarb
Watermelon mosaic virus (Potyvirus)	Aphids (non-persistent)	Cucurbits
Zucchini yellow mosaic virus (Potyvirus)	Aphids (non-persistent)	Cucurbits

Certain viruses need to multiply in the cells of the insect vector as the virus circulates before transmission can occur. This type of transmission is termed circulative propagative transmission.

Viruses can also be spread through vegetative propagation using infected plant parts (e.g. bulbs, corms, cuttings and tissue-cultured plantlets) and some are also transmitted through seed, contact or infected pollen.

Table 1.3 summarises how some important plant viruses occurring in Australia are spread.

Symptoms caused by viruses are varied and several viruses infecting one crop type may have similar symptoms, requiring laboratory tests to determine which virus is present. Symptoms of virus infection are sometimes difficult to separate from those caused by chemical damage, insect feeding and nutrient imbalances.

Viroids are smaller than viruses and are among the smallest infectious agents known. A viroid consists of a small, circular, infectious nucleic acid and is entirely dependent on the host for its reproduction. Viroids spread from plant to plant in infected propagation material and in infected sap carried on hands or on cutting and pruning instruments. Viroids occurring in Australia include avocado sun blotch, citrus exocortis, potato spindle tuber and pear blister canker.

■ NEMATODES

Nematodes are microscopic, non-segmented roundworms that belong to the animal kingdom. They occur in almost every soil and water habitat in the world and most nematode species feed on bacteria and organic matter. Nematodes attacking plants have a hollow, spear-like structure (stylet) near the mouth, used to pierce the wall of plant cells and ingest cell contents. Feeding often results in the formation of galls or lesions on roots or distortion of other plant parts. Nematodes move by swimming in films of water between soil particles or on plant surfaces. They spread by water, movement in infested soil and on contaminated machinery, and in infected planting material.

Plant-parasitic nematodes are grouped according to their feeding behaviour. **Ectoparasites** feed from the outside of the root and do not enter the plant tissues, whereas **endoparasites** enter the root or shoot tissues and feed within the plant. In addition, they may be **sedentary,** establishing specific feeding sites where they remain until they die, or **migratory,** feeding while moving through the plant tissue. Table 1.4 lists some plant parasitic nematodes affecting vegetable crops in Australia.

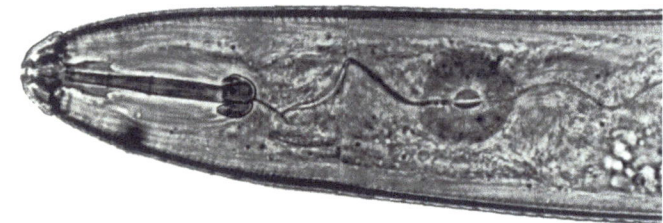

Fig 1.18 The head of a plant parasitic nematode showing the stylet, a spear-like stucture. Left: Root-knot nematode symptoms.

■ SYMPTOMS OF DISEASE

The first step in the diagnosis of a disease is recognising the visible signs or symptoms in the plant. Symptoms are the results of disturbing one or more of the vital functions of the plant, such as:

- water and mineral uptake by roots (e.g. root rots).
- carbohydrate, water and mineral translocation (e.g. vascular wilts and cankers).
- photosynthesis and respiration (e.g. leaf blights, leaf spots, mosaics).
- reproduction (e.g. fruit rots, smuts).

Most diseases produce characteristic symptoms that allow an accurate diagnosis or, alternatively, narrow it down to a few possibilities. Sometimes a definite diagnosis can be made only by using laboratory tests that allow the pathogen to be isolated from diseased tissue and identified. These tests should always be done if there is any doubt about the cause of a particular disease or if problems are being encountered during its control. Information that helps identify the cause of a disease includes the cultivar affected, location of affected plants in a field, weather conditions, crop sequences, and fertiliser and chemical treatments applied to the crop.

Table 1.4 Plant-parasitic nematodes of Australian vegetable crops

GENERA	COMMON NAME	MODE OF PARASITISM	SYMPTOMS
Meloidogyne	Root-knot nematodes	Sedentary endoparasites of roots	Plant stunting Plant wilting Root galls Root distortion
Heterodera *Globodera*	Cyst nematodes	Sedentary endoparasites of roots	Root galls/cysts Root distortion
Rotylenchulus	Reniform nematodes	Sedentary semi-endoparasites of roots	Plant stunting
Pratylenchus	Root-lesion nematodes	Migratory endoparasites of roots	Plant stunting Root lesions
Radopholus	Burrowing nematodes	Migratory endoparasites of roots	Root necrosis Plant lodging
Hemicycliophora	Sheath nematodes	Migratory ectoparasites of roots	Root tip swelling Root stunting Plant stunting
Trichodorus *Paratrichodorus*	Stubby root nematodes	Migratory ectoparasites of roots	Plant stunting Plant wilting Stubby roots
Tylenchorhynchus *Merlinius*	Stunt nematodes	Migratory ectoparasites of roots	Plant yellowing Plant stunting
Paratylenchus	Pin nematodes	Migratory ectoparasites of roots	Plant yellowing Plant stunting Root stunting
Rotylenchus *Helicotylenchus*	Spiral nematodes	Migratory ectoparasites of roots	Root lesions
Ditylenchus	Stem and bulb nematode		

■ DISEASE DEVELOPMENT AND MANAGEMENT

All diseases caused by pathogens are the result of an interaction between the host plant, a pathogen, and environmental factors such as light, temperature and moisture. Environmental factors affect the development of both the host and the pathogen. This interaction is known as the 'disease triangle' and all components must be compatible for a disease to develop (Fig 1.3).

Disease management strategies aim to favour the host plant's growth and development while attacking vulnerable stages in the life cycle of the pathogen to prevent or restrict its development. The three key means of disease management are: exclude the pathogen; reduce inoculum levels of the pathogen; and protect the host plant.

Exclusion or eradication
- Pathogen-tested seed and vegetatively propagated materials are used (e.g. budwood, cuttings and nursery trees produced under strict hygiene procedures).

- Quarantine, including international, national and State quarantine zones, prevents movement of infected plant material. Illegal movement of material is a major threat to several of Australia's horticultural industries.

- The eradication of a pathogen before it becomes widespread. This is more likely to succeed if an incursion is detected soon after it has occurred and the pathogen cannot be dispersed by air-borne spores or insect vectors.

Reduce inoculum levels
- Crop rotation reduces pathogen populations during the growth of non-susceptible crops. For soil-borne pathogens, crop rotation produces the greatest result for pathogens that survive only on living hosts or on host residues. Crop rotation is less effective for pathogens that can persist in the soil for long periods in the absence of a susceptible host.

- Incorporating organic manures into the soil increases the activity of microorganisms antagonistic to soil-borne plant pathogens.

- Sanitation includes all activities aimed at eliminating or reducing the amount of inoculum present in a plant, field or packing house. Measures include removing diseased plant material by chipping or pruning to reduce disease carryover, promptly destroying crop residues and alternative weed hosts, removing diseased fruit and thoroughly cleaning packing facilities and equipment with a suitable disinfectant. Some disinfectants are de-activated when contaminated with organic matter. Solutions should be changed regularly to ensure that disinfectant efficacy is maintained. Information about disinfectants used for sanitation in vegetable production is given in Box 1.1.

- Chemical and physical soil treatments, such as fumigation, solarisation and mulching reduce levels of soil-borne pathogens.

- Heat-treating planting material in nurseries or hydydroponic facilities, seed, bulbs and cutting is recommended. A general procedure for hot water treating seed is given in Box 1.2.

Protect the host

- Use resistant varieties. Resistance to a pathogen either prevents infection or slows disease development.

- Apply fungicides. For many vegetable diseases, targeted fungicide applications are an important component of a disease management program. Fungicides have either a protectant or systemic action (Box 1.3).

- Use defence activators. These non-pesticide agents are applied before pathogen infection to activate the plants' inherent resistance mechanisms. They may be of synthetic origin (e.g. a formulated chemical), or biological (e.g. non-pathogenic microorganisms or their products). Defence activators are used in conjunction with traditional methods of disease management.

- Apply insecticides. Insect vectors of some viruses are managed effectively with insecticides, but the mode of virus transmission is a crucial factor. Persistently transmitted viruses that require long feeding times by the vector may be controlled. However, spread of non-persistently transmitted viruses may actually be increased because vectors require feeding times of only seconds and in the short term, their feeding activity may increase after contact with insecticide.

Modern disease management aims to provide a combination of suitable methods to obtain effective, economically sound disease control with minimal risk to the environment. These principles form the basis of an integrated pest management strategy (IPM). General principles of IPM disease management are given in **Box 1.4**.

BOX 1.1 AGRICULTURAL DISINFECTANTS USED IN AUSTRALIAN VEGETABLE PRODUCTION SYSTEMS

N.B. Information is correct at the time of publication.

ACTIVE INGREDIENT	RECOMMENDED CONCENTRATION	RECOMMENDED EXPOSURE TIME	EFFICACY REDUCED BY ORGANIC MATTER?	NOTES
Farmcleanse®	10%	10 minutes	Yes	Not corrosive Biodegradable Doesn't bleach clothing
Chlorine Dioxide	500 ppm (refer to specific label instructions)	10 minutes	No	Corrosive Effective over a wide pH range (pH 6–10)
Quaternary Ammonium	0.1–1%	10 minutes	No	Slightly corrosive Effective over a wide pH range
Hydrogen peroxide + Peroxyacetic acid	0.05–1%	At least 45 seconds	Yes	Corrosive
Iodine compounds	10–30 ppm	30–60 seconds	Yes	Corrosive
Hypochlorites	0.5–2%	10 minutes	Yes	Inexpensive Corrosive Efficacy sensitive to pH More effective at pH 7–8 than at pH 10–11

BOX 1.2 HOT WATER TREATMENT FOR VEGETABLE SEED

Caution! Before beginning you should be aware:

Hot water treatment may reduce seed germination if the treatment conditions are not precisely controlled.

If you treat seed, the seed company liability and guarantees are null and void.

Coated seed or seed treated with fungicide should not be hot water treated.

Ensure that the seed has not already been hot water treated. A second treatment can kill the seed.

Before committing a large batch of seed to hot water treatment, treat a small quantity of seed first and monitor the effect on its germination.

Seed should be planted as soon as possible after hot water treatment.

Do not store treated seed for long periods as seed viability will decline.

Select the appropriate treatment conditions for the seed:

CROP	TEMPERATURE (°C)	TREATMENT TIME (MINUTES)
carrot, cauliflower, broccoli	50	20
tomato, eggplant, spinach, cabbage, brussels sprouts	50	25
celery	50	30
capsicum	51	30

You will need:

electric frypan, saucepan, nylon stocking, precision thermometer (0–60°C range, calibrated in increments of 0.5°), spoon, stopwatch or clock, absorbent paper tray.

Method:

1. Treat only small quantities of seed (up to 100 g at a time).

2. Pour the seed into the toe of a stocking and tie the top securely with a clip or a rubber band.

3. Fill the bottom one-third of the electric frypan with warm water and turn the frypan on to warm.

4. Fill the saucepan approximately two-thirds full of warm water ($\approx 50°C$) and place the saucepan in the frypan.

5. Gently stir the water in the frypan and saucepan and measure the water temperature by immersing the tip of the thermometer approximately half-way into the saucepan so that the thermometer tip is not touching the saucepan base or sides.

6. Heat the water about 0.5°C above the required temperature and ensure that this temperature is maintained – while the water is stirred – for 5 minutes before adding the seed.

7. Immerse the 'seed stocking' into the saucepan while continuing to stir the water gently.

8. Monitor the temperature and add heat if necessary by turning on the thermostat, heating for 5–10 seconds and then turning the thermostat off.

9. Keep monitoring the temperature and adding heat as required while stirring the water for the required time interval.

10. When treatment is complete, remove the stocking from the saucepan and run under cold water to remove the heat.

11. Spread the seed onto an absorbent paper tray and place in a warm location, out of direct sunlight, until the seed is dry.

12. Fungicide seed dressings may be applied before the seed is planted.

BOX 1.3 USING FUNGICIDES

Protectant fungicides act at the application site, either killing fungal spores or preventing their germination. They do not penetrate plant tissues and so do not stop disease developing if the infection has occurred prior to application. Instead, they act in a similar way to a raincoat, and are only effective if applied before infection occurs. Also called broad-spectrum fungicides or multi-site activity fungicides, they act against a wide range of fungi and fungal populations that, typically, do not develop resistance to them. They often form the basis of a fungicide spray program, and frequently as tank-mix combinations with systemic fungicides. Although protectants are cheaper than systemic fungicides, they must be applied regularly (every seven to 10 days).

Protectants are prone to weathering because they are confined to the plant surface, and good spray coverage of the target plant is essential for effective disease control.

Systemic fungicides penetrate the host tissue and control or eradicate established infections. Systemic fungicides are often more specific than protectants in the fungi they control. They are more expensive than protectant fungicides, but because they act within the host, they are less prone to weathering and incomplete spray coverage does not adversely affect their efficacy. Regular, routine use of systemic fungicides is not recommended, as it may lead to fungal populations becoming dominated by strains with fungicide resistance, reducing efficiency. Fungicides are categorised by the mechanism of fungicidal activity. Fungicides in the same group act on the same part of the metabolic pathway of the fungus. Typically, if a fungus develops resistance to one fungicide in an activity group, the other fungicides in that group will also be rendered ineffective for controlling the fungus. This phenomenon is called **cross-resistance**.

Strategies to prevent the build-up of resistant strains:

- alternate between groups of systemic fungicides

- apply protectant fungicides as the basis of the spray program and use systemics when conditions are favourable for infection

- tank-mix systemic fungicides with protectant fungicides

- use fungicides from at least two different systemic groups on the same crop

- destroy old or infected crops immediately after harvest

- when conditions favour disease development, do not wait for symptoms to develop; apply two consecutive systemic sprays from the same chemical group and then alternate with other systemics or protectant fungicides.

In Australia, fungicides must be registered for use against a particular pathogen on a particular crop. The Australian Pesticide and Veterinary Medicines Authority (APVMA) administer fungicide registration. It is illegal to apply unregistered products to vegetable crops and heavy penalties apply. Before applying fungicides, ensure that the product use is approved by checking the fungicide label.

For registered **trade names** of products and current labels, refer to the Infopest AgVet CD-ROM, which is available from Queensland Primary Industries and Fisheries (www.dpi.qld.gov.au/infopest). Information about current pesticide registrations is also available on the APVMA website: http://www.apvma.com.au. Information about current fungicide groupings and resistance management is available on the Croplife website: www.croplifeaustralia.org.au

BOX 1.4 CHECKLIST OF DISEASE MANAGEMENT PRINCIPLES FOR AN IPM STRATEGY

- Select resistant varieties.

- Source disease-free seed or transplants.

- Heat treat seed if bacterial diseases are a major problem.

- Avoid double cropping or cropping after crops in the same family.

- Use preventative fungicidal treatments when diseases are consistently a problem, otherwise apply fungicides only when conditions are conducive and symptoms are evident.

- Apply optimal water and nutrient requirements (confirm with appropriate water, soil and tissue analysis).

- Chip or remove diseased plants.

- Practise good sanitation, particularly in the presence of bacterial diseases.

- Manage virus vectors when virus is present in the crop or in a neighbouring area.

- Manage weeds within and in surrounding crops, particularly those that host diseases that affect your crop.

- Minimise mechanical damage of growing plants, and of harvested produce.

- Ensure good sanitation practices are followed when washing produce.

- Harvesting crops in cooler conditions and rapid post-harvest cooling will minimise post-harvest breakdown.

- Plough in crop residues soon after harvest.

■ FURTHER INFORMATION

General

Agrios GN (2005). *Plant pathology* (5th edn). Elsevier Academic Press, New York.

Annual Review of Phytopathology (ongoing series) http://www.annualreviews.org

Brown JF and Ogle HJ (Eds) (1997). *Plant pathogens and plant diseases*. Rockvale Publications, Armidale.

Carefoot ER and Sprott GL (1967). *Famine on the wind*. Rand McNally & Co, Chicago.

Dugan FW (2008). *Fungi in the ancient world*. APS Press, St Paul, Minnesota.

Hall IR, Brown GT and Zambonelli A (2008). *Taming the truffle: the history, lore and science of the ultimate mushroom*. APS Press, St Paul, Minnesota.

Holliday P (1998). *A dictionary of plant pathology* (2nd edn). Cambridge University Press, Cambridge.

Horst KR (2001). *Westcott's plant disease handbook* (6th edn). Kluwer Academic Publishers, Massachusetts.

Koike ST, Gladders P and Paulus AO (2007). *Vegetable diseases. A colour handbook*. Academic Press, New York.

Large EC (2003). *The Advance of the fungi*. APS Press, St Paul, Minnesota.

Madden LV, Hughes G and van den Bosch F (2007). *The study of plant disease epidemics*. APS Press, St Paul, Minnesota.

Schumann G (1991). *Plant diseases: their biology and social impact*. APS Press, St Paul, Minnesota.

Schumann GL and D'Arcy CJ (2006). *Essential plant pathology*. APS Press, St Paul, Minnesota.

Sherf AF, Macnab AA (1986). *Vegetable diseases and their control* (2nd edn). John Wiley and Sons Ltd, Australia.

Shivas RG and Hyde KD (1996). Biodiversity of plant pathogenic fungi in the tropics. In *Biodiversity of tropical fungi*. (Ed. KD Hyde) pp. 47–62. University of Hong Kong Press, Hong Kong.

Snowden AL (1990). *A colour atlas of post-harvest diseases and disorders of fruits and vegetables*: Volume 1 General introduction and fruits. Wolfe Scientific, London.

Waller JM, Lenne JM and Waller SJ (Eds) (2002). *Plant pathologists pocketbook* (3rd edn). CABI Publishing, Wallingford.

The American Phytopathological Society http://www.apsnet.org

The Australasian Plant Pathology Society. http://www.australasianplantpathologysociety.org.au

Plant Health Australia http://www.planthealthaustralia.com.au

Bacterial diseases

Fahy PC and Persley GJ (1983) *Plant bacterial diseases, a diagnostic guide*. Academic Press, Australia.

Streten C and Gibb K (2006) Phytoplasma diseases in sub-tropical and tropical Australia. *Australasian Plant Pathology* **35**, 129–146.

Fungi and fungal diseases

Dugan F (2006). *The identification of fungi: an illustrated introduction*. APS Press, St Paul, Minnesota.

Holliday P (1980). *Fungus diseases of tropical crops*. Cambridge University Press, Cambridge.

Kirk PM, Cannon PF, David JC and Staplers JA (Eds) (2001). *Ainsworth and Bisby's dictionary of the fungi* (9th edn). CABI Publishing, Wallingford.

Nematode diseases

Bridge J and Starr J (2007). *Plant nematodes of agricultural importance – a colour handbook*. Manson Publishing, United Kingdom.

O'Brien PC and Stirling GR (1991). *Plant nematology for practical agriculturalists* (3rd edn). Department of Primary Industries, Brisbane, Queensland.

Shurtleff MC and Averre CW (2000). *Diagnosing plant diseases caused by nematodes*. APS Press, St Paul, Minnesota.

Stirling GR, Harrower K and Webb LE (Eds) (2008). Plant and soil nematology in Australia and New Zealand. *Australasian Plant Pathology* **37**, 3 (special issue).

Virus diseases

Hadidi A, Khetarpal RK and Koganezawa H (Eds) (1998). *Plant virus disease control*. APS Press, St Paul, Minnesota.

Hull R (2002) *Matthews' plant virology* (4th edn). Academic Press, New York.

Loebenstein G and Thottappilly G (Eds) (2003). *Virus and virus-like diseases in major crops in developing countries*. Kluwer Academic Publishers, London.

Walkey D (1991). *Applied plant virology* (2nd edn). Chapman and Hall, London.

COMMON DISEASES OF VEGETABLE CROPS

■ BACTERIAL SOFT ROTS

Cause
Three species belonging to the *Erwinia* group of bacteria can cause soft rots. *E. carotovora* subsp. *carotovora* is the most common and widely distributed soft rot bacterium. *E. carotovora* subsp. *atroseptica* is a low temperature variant that is primarily restricted to potatoes, and *E. chrysanthemi* causes most damage at temperatures above 30°C.

Importance and hosts
Bacterial soft rots affect many vegetables, attacking the fleshy storage organs, fruit and succulent stems or leaves. Although plants can be affected in the field, soft rots are generally a problem during transport and storage. Vegetables commonly affected include carrot, cucurbits, potato, cabbage, celery, capsicum, lettuce, ginger and tomato.

Symptoms
The symptoms of soft rots are similar on all hosts. Small, water-soaked lesions appear and rapidly enlarge in diameter and depth, becoming soft and mushy. Bacterial slime often covers affected areas and rotting tissues produce a foul smell. The fruit flesh rots, rather than the skin, which may remain as a hollow shell containing an unpleasant smelling liquid.

Fig 2.2 Bacterial soft rot in carrots.

Source of infection and spread
The bacteria survive in decaying vegetable matter and crop debris in the soil. Infection occurs through damaged tissue, and soft rots often develop following injuries from mechanical damage, insects or other diseases. Hot, wet weather favours rapid development of soft rots. In the field, spread occurs by water splash, contaminated cutting knives, and insects. During transit and storage, soft rots spread by contact, in bacterial ooze dripping from diseased leaves and in contaminated washing water.

Fig 2.1 Bacterial soft rot in asparagus stems.

Fig 2.3 Bacterial soft rot causing a wet, slimy rot in celery leaves and stalks.

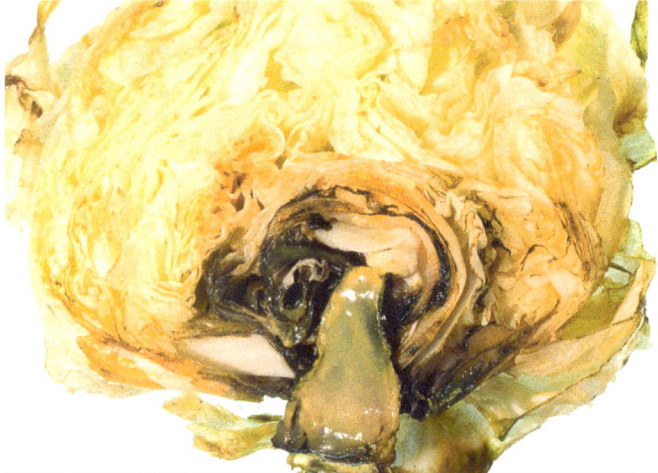

Fig 2.5 Bacterial soft rot in Chinese cabbage.

Management

The control of bacterial soft rots in vegetables depends, largely, on good sanitation and cultural practices.

- Avoid harvesting while plants are wet.

- Handle produce carefully to minimise wounding.

- Clean and disinfect harvesting implements, packing sheds and washing equipment.

- Cool produce at recommended temperatures after harvest and during transport to market.

- Apply post-harvest dip treatments if recommended for a particular crop.

- If washing fruit, properly sanitise and change the water frequently to prevent contamination by bacteria.

Fig 2.4 Bacterial soft rot in lettuce. Bacterial ooze can be seen on the cut lettuce.

LIBERIBACTER DISEASES IN TOMATO AND RELATED CROPS – BIOSECURITY THREAT

Cause

A bacterium *Candidatus* Liberibacter solanacearum.

Symptoms

Affected tomato plants are stunted, with leaf curling and yellowing. The leaf axil or leaf stalk may also become very long. Fruit may be misshapen, with a strawberry-like appearance.

Symptoms in capsicum can include pale green or yellow leaves with spiky tips and shortened leaf stalks. Flower drop may occur and parts of the plant may die back.

Infected potato plants are stunted with small yellow, often curled leaves.

Fig 2.6 Leaf curling, yellowing and stunting in a tomato plant infected by Liberibacter.

Fig 2.7 Liberibacter infection in a glasshouse-grown tomato plant. Leaf stalks are longer than usual and the plant is pale and yellow compared with healthy neighbours.

Source of infection and spread

The pathogen is a bacterium limited to the phloem tissues of plants and is distinct from other species of *Liberibacter* causing disease in citrus and other hosts.

Fig 2.8 Liberibacter infection of capsicum. Note the pale green-yellow leaves with spiky tips. Inset: affected terminal growth.

Fig 2.9 Liberibacter symptoms on potato. The plant is stunted with small, chlorotic, curled leaves. Inset: Discolouration inside the tuber.

The organism is spread by the tomato/potato psyllid *Bactericeria cockerelli*.

In addition to tomato and capsicum, the bacterium has been found on potato, tamarillo, Cape gooseberry and chilli, all members of the potato or Solanaceae family.

Importance

The pathogen was identified from glasshouse-grown tomato and capsicum plants in New Zealand in 2008. The psyllid vector was first found in New Zealand in 2006.

What to do if you suspect Liberibacter disease

This pathogen is a biosecurity risk to Australia. Any suspected affected plants should be reported to the nearest Department of Primary Industries or the Plant Health Australia hotline (1800 084 881).

■ BOTRYTIS DISEASES

Cause

The fungi *Botrytis* spp.

Importance and hosts

Diseases caused by species of the genus *Botrytis*, particularly *B. cinerea*, are among the most common and widely distributed diseases of fruit, vegetable and ornamental crops. The fungus is a very common cause of flower blight and fruit rot, but it may also cause damping-off of seedlings, leaf spots and rots of vegetative plant parts such as bulbs, tubers and corms. Some common diseases caused by *Botrytis* species are grey mould rot of tomato, bean, capsicum, cucumber, brassicas and lettuce, neck rot of onion, and chocolate leaf spot of faba bean. *Botrytis* also causes secondary rots of fruits and vegetables in storage, transit and in the market place. The fungus can also be a common cause of disease in greenhouse-grown crops.

Fig 2.10 Botrytis rot on Lebanese cucumber. The infected area is covered with grey fungal hyphae, and also shows an advanced water-soaked lesion.

Symptoms

Flower petals are very susceptible to *Botrytis* infection, particularly as they age. Blossom blights often precede and lead to fruit rots, which typically begin as a blossom-end rot. Affected fruit become water-soaked and soft, and

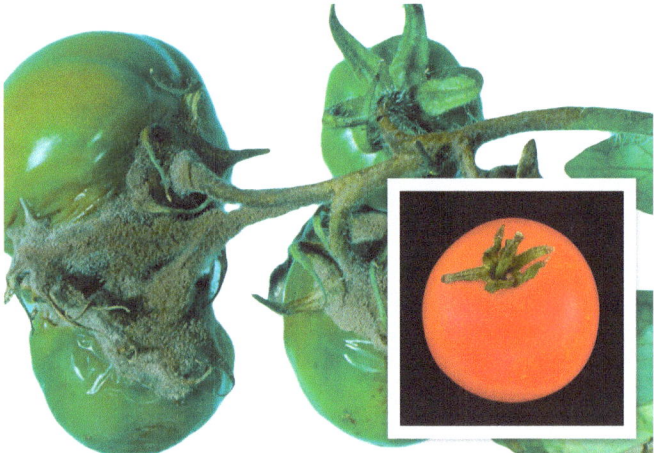

Fig 2.11 Botrytis rot on tomato. The fungus often invades floral parts, and then infects developing fruit. Inset: ghost spot symptoms.

Fig 2.12 Botrytis rot on harvested zucchini fruit.

rapidly covered by an abundant grey mould. Plant stems may also become blighted if they contact infected blossoms. *Botrytis* leaf spot of faba and broad beans causes grey to brown leaf spots that may coalesce to blight leaves.

Source of infection and spread

Botrytis has a very wide range of hosts. Spores disperse in wind and enormous numbers are produced on furry mould on infected plants. The fungus can survive from season to season as sclerotia in the soil or in crop residues. *Botrytis* is a weak parasite, although the fungus can penetrate plant tissues directly through the formation of appressoria if provided with a food source. Grey mould infection often begins when the fungus establishes in senescing floral parts, producing abundant spores during cool, humid weather.

Cool weather with heavy dew or fog favours the disease. Showery weather is not necessary for severe outbreaks.

Fig 2.13 Botrytis leaf spot (chocolate leaf spot) on faba and broad beans.

Management

- Destroy crop residues after harvest and remove diseased fruit from packing sheds.

- Reduce humidity in greenhouses and shade houses by ventilation and heating.

- Apply registered fungicides as field sprays or post-harvest treatments.

■ DAMPING-OFF DISEASES OF SEEDLINGS

Cause

Damping-off is the destruction of seedlings by pathogens. The pathogens most commonly responsible for damping-off are *Pythium* species and *Rhizoctonia solani*. Other pathogens that may be involved are *Phytophthora*, *Fusarium* and *Aphanomyces*. Many other fungi and some bacteria can cause damping-off symptoms when carried in or on the seed.

Fig 2.15 Damping-off in silver beet seedlings in the field.

Importance and hosts

Pythium spp. and *Rhizoctonia solani* are extremely common in soils in both tropical and temperate regions. Seedling plants of almost all fruit, vegetable, field and ornamental crops are liable to attack, with soil and weather conditions playing a major role in development of the disease. Plants can be attacked following direct sowing in the field, in outside seedbeds and in glasshouse seedling-production areas.

Symptoms

There are two main types of damping-off. Pre-emergence damping-off results in the rot of seed or seedling sprouts before emergence.

Fig 2.14 Damping-off in celery seedlings, causing stunting and plant death.

Fig 2.16 Seedlings in the centre show symptoms of damping-off caused by *Pythium*.

Post-emergence damping-off occurs after plants have emerged from the soil. A soft decay of the taproot or rootlets causes collapse of seedlings. Infection by *R. solani* results in a water-soaked, sunken lesion at ground level, causing the plant to fall over. Surviving plants are stunted, and affected areas often show uneven growth.

Seedlings with more rigid stems (e.g. cabbages) do not fall over, but the stem is thin, discoloured and may be bent or twisted without breaking, giving the disease the name 'wire stem'. Eventually, the stems are girdled and the plant dies.

Source of infection and spread

Pythium species are widely distributed in soil and water, living as saprophytes or as parasites attacking fibrous roots of plants. Different species are adapted to different temperature regimes. The pathogens thrive under wet conditions, and seed or seedlings growing in wet soil are most likely to be attacked, particularly when temperatures are unfavourable for the host.

Rhizoctonia solani is very common in soil, surviving as sclerotia or mycelium. The fungus survives between crops on plant trash. The plant residues provide a ready source of inoculum from which disease can develop in subsequent crops. Moderately wet rather than waterlogged soil favours this fungus. Infection is more likely in plants growing poorly, whereas rapidly growing, vigorous plants often escape infection.

Both *Pythium* and *R. solani* are spread by rain, irrigation, water splash, contaminated tools, soil or potting mix, and infected plants.

Management

- Raise seedlings in fumigated soil or soil-less potting mix on elevated benches.

- Prevent contamination of treated soil by avoiding water splash, contaminated tools or storage in dirty areas.

- Treat water supplies before use if dams are the source of water.

- Avoid excessive watering, and thin seedlings to allow good air circulation.

- Treat seed with the recommended fungicides before planting and apply as necessary if outbreaks of damping-off occur.

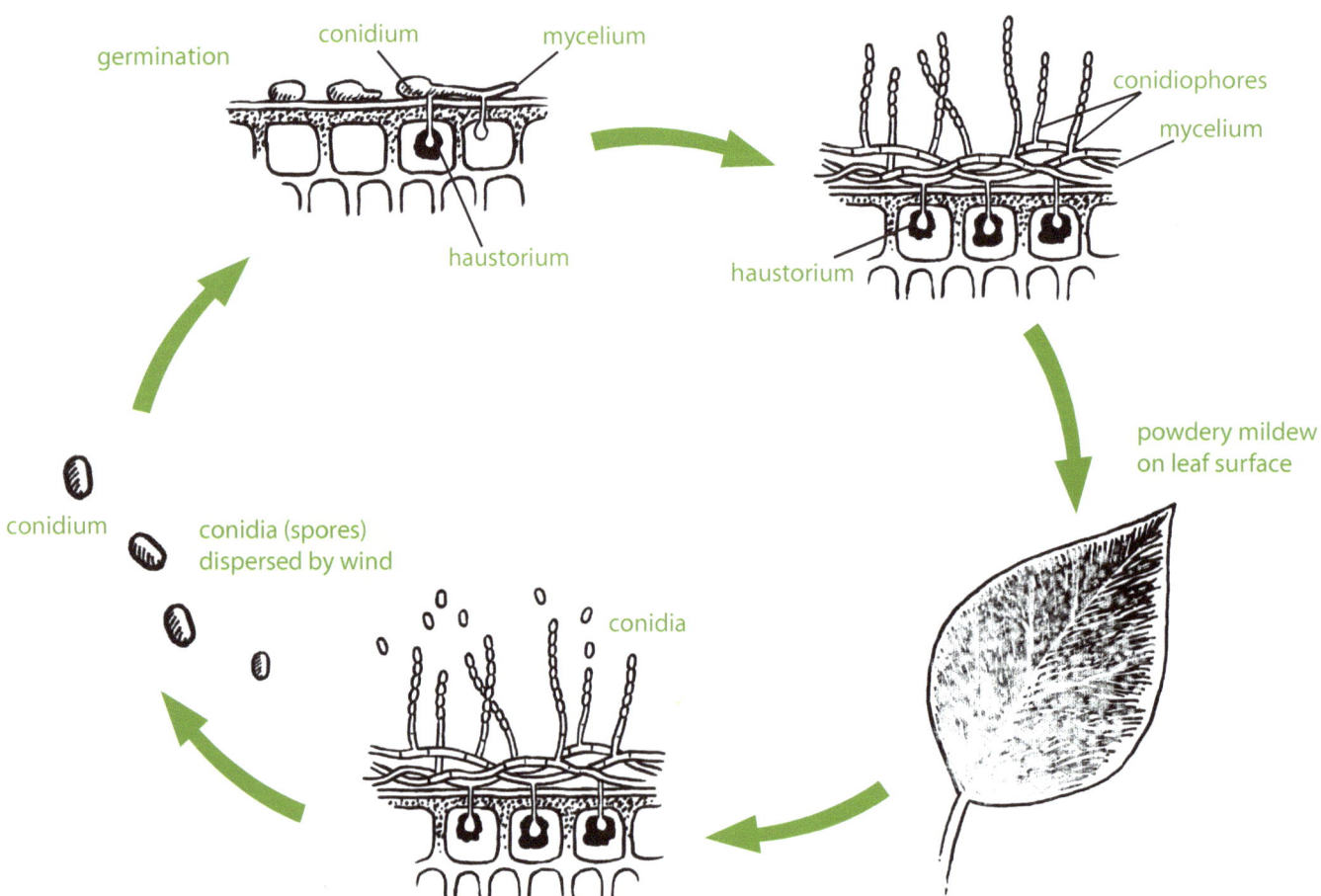

Fig 2.17 Disease cycle of powdery mildew.

■ POWDERY MILDEWS

Powdery mildew is probably the most common of all plant diseases. The characteristic white, powdery growth occurs on plants as diverse as cereals, trees, turf grass, woody ornamentals and most vegetable and fruit crops.

Cause

Although the symptoms of powdery mildew are similar on many hosts, several fungal species cause the disease. Many are host specific, often infecting only a few, related species. The main genera of fungi causing powdery mildew diseases include *Erysiphe*, *Leveillula*, *Oidium*, *Podosphaera*, *Sphaerotheca* and *Uncinula*.

Races or strains may develop within a species in response to using a host-resistance gene for management.

Symptoms

Powdery mildew appears as spots or patches of white to greyish, powdery growth (mycelium) on the surface of leaves and other plant parts. The mycelium is most visible on the upper leaf surface, often covering it completely as the disease progresses.

Source of infection and spread

Fungi causing powdery mildew grow largely on the surface of plants. They are obligate parasites and obtain nutrients by sending feeding organs (haustoria) into the epidermal cells of plants (Fig 2.17).

The superficial fungal mycelium produces chains of spores (conidia) that are widely dispersed by air currents. The spores do not require free water for germination and germinate freely in relatively low humidity, including moisture from morning dews, fog or condensate. Disease

Fig 2.19 Powdery mildew on potato leaves and stems.

development is favoured by warm, dry and especially, cloudy conditions, which limit damage to the fungus by ultraviolet radiation. Humid, wet weather slows disease progress.

Overwintering fruiting bodies, called cleistothecia, may develop late in the season, or, when conditions become unfavourable. These appear as tiny, pin-head size, yellow-gold and later brown to black bodies within the mycelium. These fruiting bodies survive in leaf litter and crevices of plants or on alternative host species until spring, when ascospores are released to begin new infections. In warmer areas, the fungi may survive on alternative hosts or as mycelium or conidia in buds and other plant parts.

Fig 2.18 Powdery mildew on carrot leaves.

Fig 2.20 Powdery mildew on tomato.

Importance

Damage from powdery mildew may take some time to develop. Efficiency is reduced in affected leaves and fruit can be scarred and damaged, causing produce to be downgraded. Severe outbreaks can cause defoliation, exposing fruit to sunburn and predisposing them to secondary rots.

Management

- Use appropriate cultural management procedures, including removal of diseased twigs and crop debris, to reduce inoculum levels.

- Apply pre-infection (protectant and post-infection (eradicant or curative) fungicides.

- Plant resistant varieties when appropriate, and available.

Refer to specific crop sections for further information.

■ RHIZOCTONIA DISEASES

Cause

The fungus *Rhizoctonia solani*.

Importance and hosts

Rhizoctonia solani is common in most soils and is able to cause some form of disease in almost all cultivated plants. Different 'strains' of *R. solani*, called anastomosis groups, have been recognised. The anastomosis groups differ in their host range and pathogenicity. The most common diseases are damping-off of seedlings, root and stem rots, stem cankers and fruit rot. The diseases often have common names that describe symptoms on the particular host, for example base rot of lettuce, black scurf of potato tubers, crater rot of beetroot, and wire stem of tomato and crucifer seedlings.

Symptoms

Symptoms vary according to the host and the plant part affected. Damping-off of seedlings is probably the most common disease caused by *R. solani* and is discussed above.

Infection of stems causes dark-coloured cankers or rotting at the base of the stem. On plants growing close to the ground, e.g. lettuce, the fungus attacks the older leaves in contact with the soil, causing brown lesions on the leafstalks and leaves. The disease may progress to involve the whole plant, causing wilting and death.

Fig 2.21 Rhizoctonia crown rot of silver beet. Note dark cankers at the base of the plant.

Fig 2.22 Rhizoctonia rot on beetroot. Sunken lesions are often covered by greyish fungal growth.

Fig 2.23 Rhizoctonia rot on bean seedlings, showing sunken, brown lesions on lower stems.

On fleshy, succulent stems, roots and storage organs, the fungus causes brown, rotten areas or sunken cankers that may be covered by fungal mycelium. Infection of potato tubers causes black scurf in which small, dark, flat sclerotia resembling specks of dirt occur on the tuber surface and which are not removed by washing.

Fruit growing near the ground can be infected, developing firm, water-soaked areas that became sunken and often crack open.

Source of infection and spread

Rhizoctonia solani is present in most soils and, once introduced, remains there indefinitely. Its distribution in soils is often patchy, but it is found predominantly in the upper 15 to 20 cm of the soil profile as mycelium or sclerotia, or in organic debris. The anastomosis groups differ in host range, types of diseases caused and the ability to compete and survive in soil. The anastomosis groups present in a soil are strongly influenced by previous

Fig 2.24 Zonate lesions on leaves caused by *Rhizoctonia solani*.

cropping history. Spread of *R. solani* occurs with soil movement, in water, and on contaminated tools and plant parts. Rhizoctonia diseases are more severe in soils that are moderately wet rather than waterlogged or dry.

Management

- Grow seedlings in soil-less potting mixes or in sterilised or pasteurised soil.

- Treat seed with the recommended fungicide before planting.

- Maintain optimum plant growth and avoid injuring plants, as wounds provide a means of entry for the fungus.

- Where possible, avoid soil contact of fruit by using plastic mulch or by staking plants.

- Ensure crop residues decompose thoroughly before replanting an area.

- Applications of fungicides to the soil provide control of some diseases caused by *R. solani*.

■ SCLEROTINIA ROTS

Cause

The fungi *Sclerotinia sclerotiorum* and *Sclerotinia minor*.

Importance and hosts

Fungi of the genus *Sclerotinia* attack a wide range of vegetable, fruit and field crops. Two species, *S. sclerotiorum* and *S. minor*, cause important diseases in vegetables. Both species have wide host ranges. *Sclerotinia sclerotiorum* occurs on more than 400 different hosts, including numerous cultivated vegetables. *Sclerotinia minor* occurs on more than 90 hosts, including peanut, lettuce and potato. Diseases caused by *S. sclerotiorum* are known by several names, including white mould or nest of bean, drop of lettuce and sclerotinia rot of cabbage, carrot, celery, potato and tomato.

Symptoms

The two species of *Sclerotinia* cause similar symptoms. These include water-soaked rotting of leaves, stems or fruit, with a white, fluffy fungal mycelium covering affected areas. Compact resting bodies or sclerotia of the fungus soon develop on the rotted tissues. The sclerotia are white at first, but later become black and hard. Sclerotia of *S. sclerotiorum* are large and elongated (2–20 mm long and 3–7 mm wide), whereas those of *S. minor* are smaller and irregular to roughly spherical (0.5–2 mm in diameter).

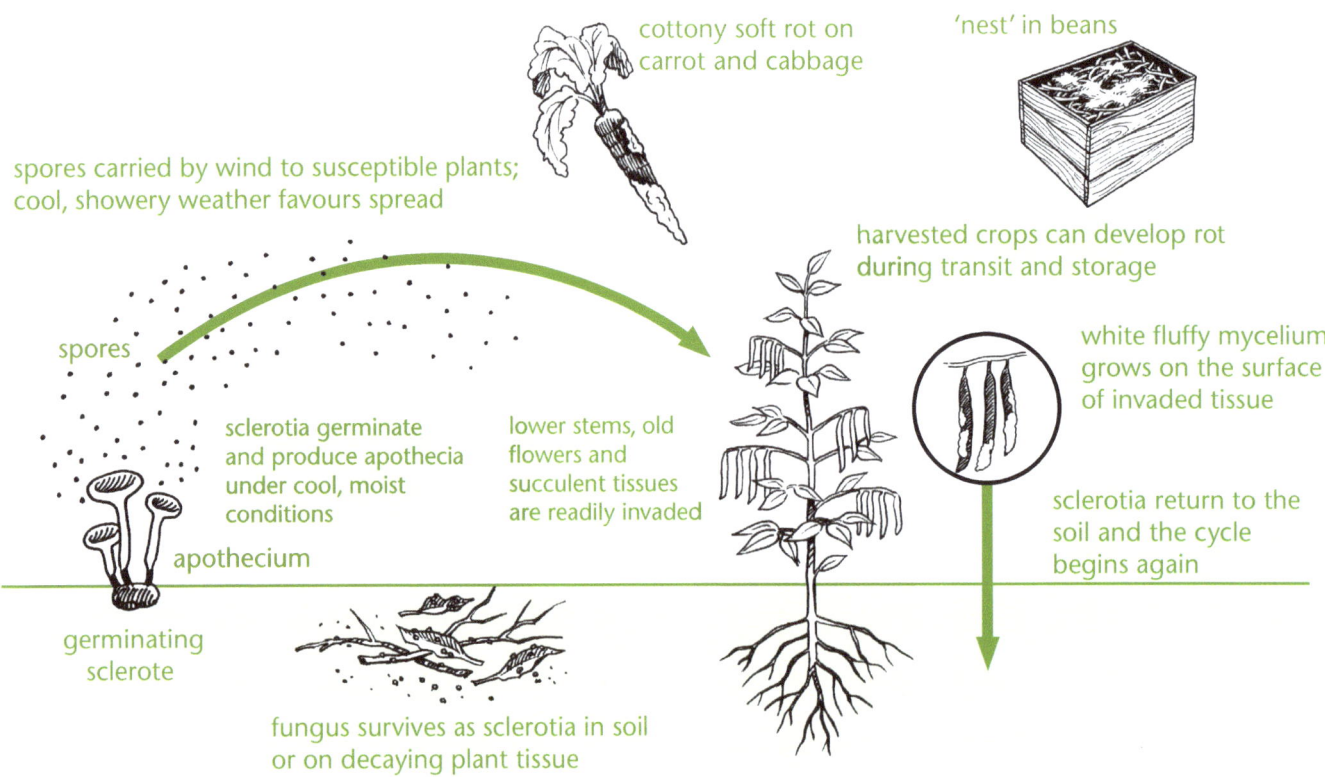

cottony soft rot on
carrot and cabbage

'nest' in beans

spores carried by wind to susceptible plants;
cool, showery weather favours spread

harvested crops can develop rot
during transit and storage

spores

white fluffy mycelium
grows on the surface
of invaded tissue

sclerotia germinate
and produce apothecia
under cool, moist
conditions

lower stems, old
flowers and
succulent tissues
are readily invaded

sclerotia return to the
soil and the cycle
begins again

apothecium

germinating
sclerote

fungus survives as sclerotia in soil
or on decaying plant tissue

Fig 2.25 Disease cycle of *Sclerotinia sclerotiorum*.

Stem infection by *Sclerotinia* causes a pale or dark brown lesion, often quickly covered by white cottony patches of fungus. Leaves may show little evidence of attack, but the fungus can rapidly rot the stem, causing collapse of the plant.

Leaves and petioles of plants such as celery and lettuce rapidly wilt and die following infection, with total collapse of the plant occurring as the fungus spreads through the stem.

Source of infection and spread

Sclerotinia species have a wide range of hosts among crop, weed and ornamental species. Sclerotia formed in infected plants can survive in the soil for many years.

Sclerotinia sclerotiorum: During moist weather, sclerotia of *S. sclerotiorum* near the soil surface germinate and produce slender stalks that terminate in small, cup-shaped structures called apothecia. Ascospores produced in the apothecia discharge into the air and are carried by wind. Ascospores germinate when they land on senescing plant parts such as old blossoms and leaves, which provide a readily available food source. The fungus then multiplies and rapidly infects adjacent healthy tissue. Ascospore

germination occurs in free water from dew, rain and fog or sprinkler irrigation.

Sclerotinia minor: A senescing food base is not required for *S. minor* infection. Instead, sclerotia germinate to produce fungal mycelium that infects plants directly.

Fig 2.26 Sclerotinia rot in potatoes showing collapsed plants and a healthy row on the left.

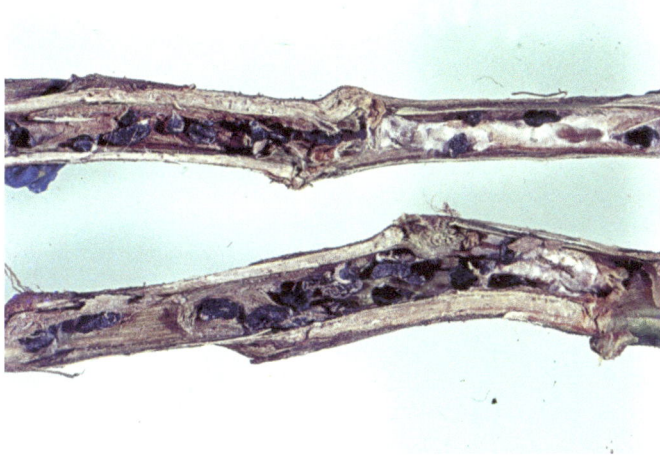

Fig 2.27 Sclerotinia rot in tomato. The split stems show white fungal growth and large, black sclerotia (rusting bodies).

Cool, wet weather favours sclerotinia rots. Disease can occur at temperatures ranging between 4–30°C, with temperatures slightly below 20°C being optimal for infection.

Management

- Rotate susceptible crops with resistant ones such as summer and winter cereals, onion and sweetpotato.

- Plough in diseased crops immediately after harvest.

- Apply the recommended fungicides. Correct timing and good penetration of foliage are essential for effective control.

- Further information on disease management is given in specific crop chapters.

Fig 2.28 Sclerotinia rot in lettuce causing major crop losses.

Fig 2.29 Apothecia of *Sclerotinia sclerotiorum*. Sclerotes germinate to produce apothecia on slender stalks, terminating in dish-shaped apothecia, which produce asci and ascospores.

Fig 2.30 Symptoms of *Sclerotinia sclerotiorum* in cabbage.

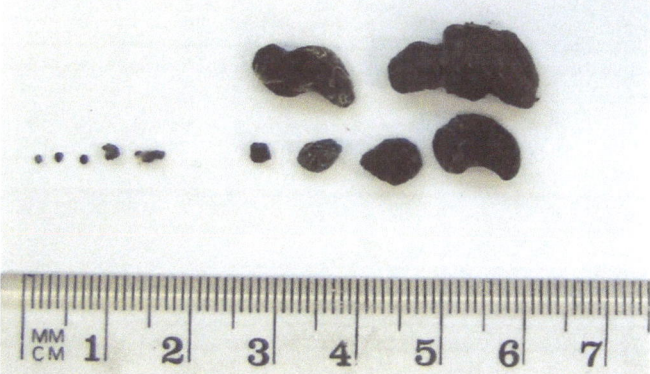

Fig 2.31 A size comparison between sclerotes of *Sclerotinia sclerotiorum* (right) and *S. minor*.

■ SCLEROTIUM DISEASES

Cause
The fungus *Sclerotium rolfsii.*

Importance and hosts
Sclerotium rolfsii is a very common soil-borne fungus infecting a wide range of vegetable, ornamental and field crops. It is most active during warm, wet weather in tropical and subtropical regions. Vegetable crops commonly affected include bean, beetroot, capsicum, carrot, cucurbits, potato and tomato. The fungus causes rots of the lower stem, roots and crown. It can also cause rot of fruit in contact with the soil.

Symptoms
Symptoms develop on plant parts in or near the soil. The most common symptom is a brown to black rot of the stem near the soil line. The stem becomes girdled and the plant wilts suddenly and dies. A coarse, white, cottony fungal growth, containing white, spherical resting bodies (sclerotia) covers the affected area. The sclerotia soon become light brown and resemble cabbage seed. Fruit symptoms usually develop where there has been contact with the soil. Decay may progress rapidly, eventually causing complete collapse.

Source of infection and spread
The fungus can survive for years as sclerotia in the soil or in host plant debris. Sclerotia spread with soil movement, infested plant material and contaminated equipment. Infection and disease development are favoured by warm, moist conditions. Sclerotium diseases often develop on crops produced under sub-optimal growing conditions, when plant vigour and quality has been compromised by other factors.

Fig 2.33 Sclerotium fruit rot on tomato showing sclerotia on fruit.

Fig 2.34 Sclerotium base rot on tomato. White mycelium usually develops near the soil line.

Fig 2.32 Detail of sclerotia on host tissue.

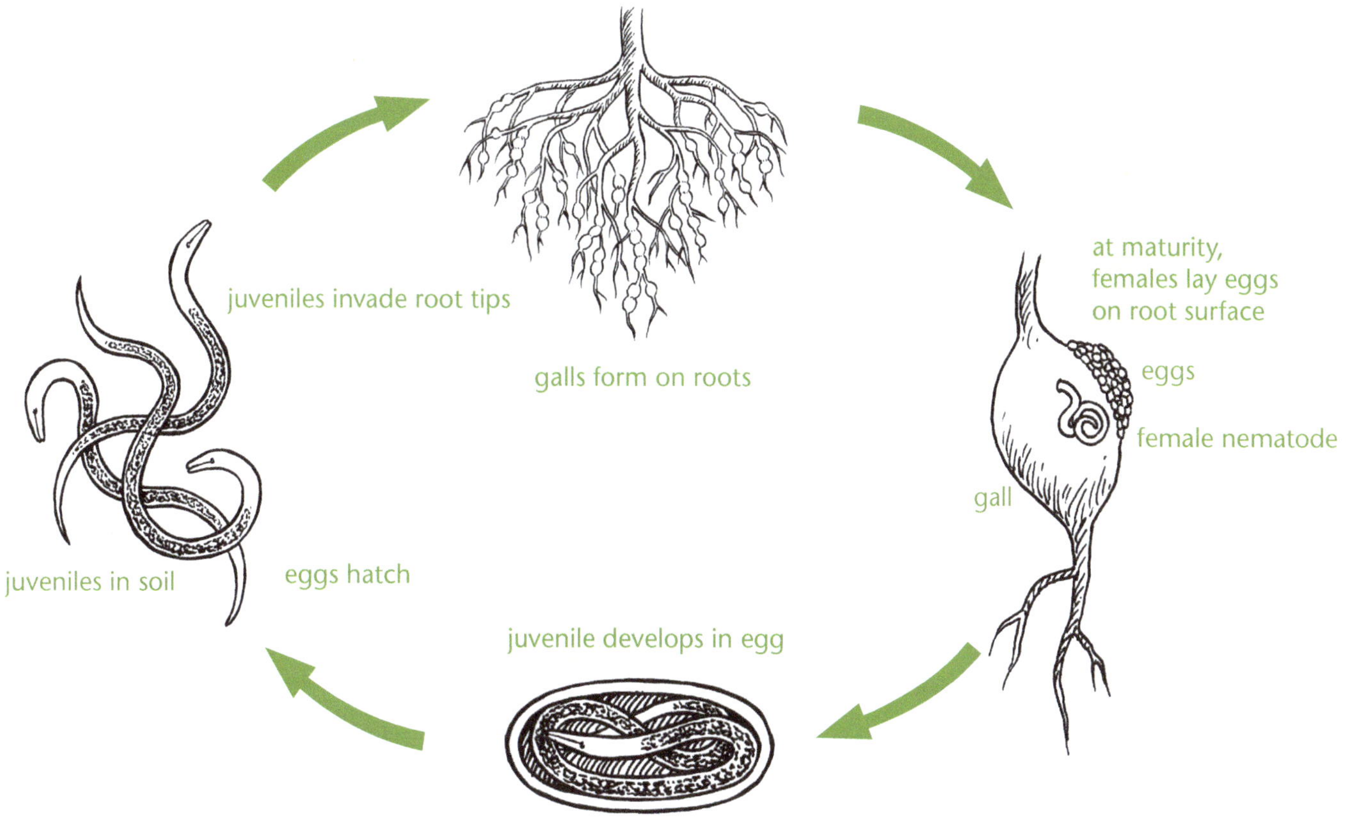

Fig 2.35 Life cycle of root-knot nematode.

The diagram labels read: juveniles invade root tips; galls form on roots; at maturity, females lay eggs on root surface; eggs; female nematode; gall; juvenile develops in egg; juveniles in soil; eggs hatch.

Management

Control of sclerotium diseases is difficult when soil and weather conditions favour the fungus. Management systems that can reduce the disease severity include the following:

- Ensure plant residues have decomposed before planting.

- Deep ploughing soil to bury host debris and sclerotia is a useful measure.

- Include non-susceptible crops such as maize and small grains in rotations to reduce inoculum levels in soil.

- Drench transplants with the recommended fungicides.

■ ROOT-KNOT NEMATODES

Cause

Root-knot nematode (*Meloidogyne* spp.). Three species (*M. javanica, M. incognita* and *M. arenaria*) are widespread, occurring throughout northern Australia and in soils under irrigation in southern States. *Meloidogyne hapla* predominates in Tasmania, coastal southern Australia and areas that are cool due to their high elevation. Other species (e.g. *M. fallax* and *M. trifoliophila*) may occur in local situations or on particular crops.

Importance and hosts

Root-knot nematodes are particularly important in regions with warm to hot climates and mild winters. Damage is most severe in sandy soils and in well-structured clay loam soils of volcanic origin.

The common species of root-knot nematodes all have a wide host range and most plants are able to host at least one

Fig 2.36 Nematode galling on carrot.

Fig 2.37 Root-knot nematode on tomato roots.

Fig 2.39 Plants wilting in the field (foreground) due to nematode root damage.

species. In general, members of the grass family (Poaceae) are less susceptible than other plants. Important hosts include:

- *Vegetables:* bean, brassicas, capsicum, carrot, celery, cucurbits, eggplant, ginger, lettuce, okra, potato, snake bean, sweetpotato, tomato.

- *Fruit crops:* almond, banana, grape, kiwi fruit, papaya, passionfruit, peach, pineapple, strawberry.

- *Field crops:* mungbean, soybean, sugarcane.

- *Pastures:* lucerne, subterranean clover, white clover.

- *Ornamentals:* chrysanthemum, carnation, rose.

Symptoms and damage

Plants affected by root-knot nematodes have an unthrifty appearance and often show symptoms of stunting, wilting or chlorosis (yellowing). Symptoms are particularly severe

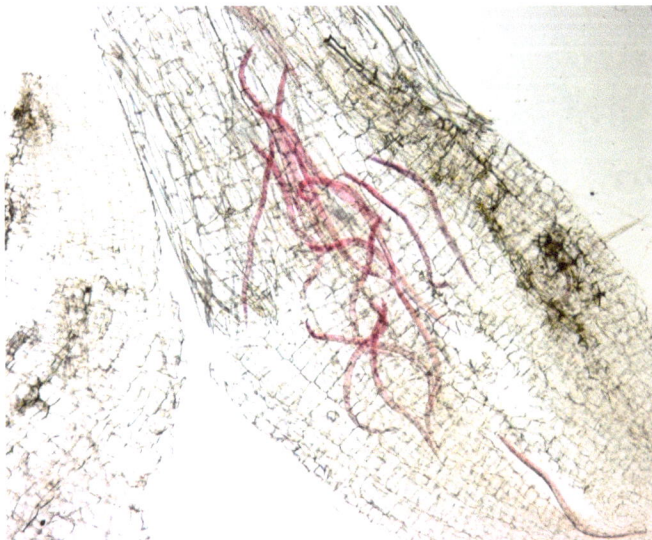

Fig 2.38 Plant parasitic nematodes inside a plant root.

when crops are planted into soil with high numbers of nematodes, as the plant is damaged during its establishment phase. More commonly, however, nematode populations increase gradually during the season, with plants growing normally until maturity. Plants then begin to wilt and die back, and fruit set and fruit development is retarded.

Below ground, the symptoms caused by root-knot nematodes are quite distinctive. Lumps or galls, ranging in size 1–10 mm in diameter, develop all over the roots. In severe infestations, heavily galled roots may rot away, leaving a poor root system with a few large galls.

Biology of root-knot nematodes

Root-knot nematode juveniles are active, thread-like worms about 0.5 mm long, too small to be visible with the naked eye. These juveniles hatch from eggs, move through the soil and invade roots near the root tip. They occasionally develop into males, but usually become sphere-shaped females. The presence of developing nematodes in the root stimulates the surrounding tissues to enlarge and produce the galls that are typical of infection by this nematode. Mature female nematodes then lay hundreds of eggs on the root surface, and these eggs hatch in warm, moist soil to continue the life cycle.

Continued infection of galled tissue by second and later generations of nematodes causes the large galls that are sometimes visible on plants such as tomato at the end of the growing season. The length of the life cycle depends on temperature and varies from four to six weeks in summer and from 10–15 weeks in winter. Nematode multiplication and the degree of damage are greatest in crops during the warmest months of the year.

Soil moisture conditions that are optimum for plant growth are ideal for the development of root-knot nematode. Nematodes cannot move unless there is a water film around soil particles and eggs will not hatch unless soil moisture is adequate.

Management

Nematode monitoring. The best way to make informed decisions about nematode management is to collect soil samples at appropriate times of the year and have them processed in a nematology laboratory. Nematode counts provide information on the effectiveness of management practices and used to predict the likelihood of damage in following crops. Monitoring the level of galling on roots is also a useful management tool.

Crop rotation. Root-knot problems increase and control becomes more difficult when susceptible crops are grown in succession. An appropriate rotation program will significantly reduce losses, but will not eliminate nematode infestations (because some eggs may survive for at least 12 months and most weeds can host the nematode). Winter cereals are useful rotation crops because they are generally poor hosts of root-knot nematode and nematode reproduction is limited by temperature during the cold winter months. However, the crop grown during summer is the most important component of a rotation program. Sorghum × sudan grass hybrids, signal grass, green panic, Rhodes grass and peanut are some summer crops with good resistance to most populations of the nematode.

Fallowing. Provided the soil is warm and moist, a weed-free fallow of three to six months will usually reduce root-knot nematode populations by more than 95%. However, the disadvantage of fallowing is that it has a negative impact on soil health and exposes soil to the risk of erosion.

Sanitation. In crops established from seedlings, it is essential that transplants are free of root-knot nematodes. Potting mixes used for seedling production will be nematode-free if the peat, sand and other components of the mix come from sources free of the nematode and the ingredients remain uncontaminated before use.

Resistant varieties. Varieties with resistance to root-knot nematodes are available in some vegetable crops (e.g. tomato and sweetpotato), and nematode-resistant rootstocks are commonly used for perennial cops such as grape and peach. Resistant varieties and rootstocks provide adequate, but not absolute, protection against most common populations of root-knot nematode.

Soil health. Plants grown in soils with good physical, chemical and biological properties are better able to withstand nematode attack than plants growing in degraded, infertile soils. The best way of restoring soil health is to increase levels of soil organic matter, as this will improve a wide range of important soil properties including physical structure, water holding capacity, cation-exchange capacity, nutrient storage and nutrient release. Since organic matter is also a food source for many parasites and predators of nematodes, root-knot nematode populations tend to decline as levels of soil organic matter increase. Organic amendments, organic inputs from rotation crops, residue retention and reduced tillage are some options for improving soil health and reducing the impact of root-knot nematode.

Integrated pest management (IPM). When used on their own, the control practices listed above will have some impact on losses from root-knot nematode. However, an IPM approach (using several control practices together) is likely to have an even greater impact. In situations where root-knot nematode is a major factor limiting crop production, the aim should be to integrate several of the above practices into the farming system.

Nematicides. When IPM is used to manage root-knot nematode, a nematicide should only be needed for chronic nematode infestations. Even in these situations, good management will reduce the nematode population pressure, so that nematicides have a greater chance of giving effective control.

■ ALFALFA (LUCERNE) MOSAIC

Cause

Alfalfa mosaic virus (Alfamovirus).

Importance and hosts

Alfalfa mosaic virus occurs worldwide and has a wide range of hosts among legumes and other broad-leafed plants. Natural hosts of the virus include lucerne, clover, potato (calico disease), tomato, capsicum, lettuce, soybean, bean, sowthistle (*Sonchus oleraceus*) and nightshades (*Solanum* spp.).

Symptoms

Symptoms can vary depending on the host, but nearly always involve a yellow-green mosaic pattern or bright yellow blotches on leaves. As the plant matures, these areas may become more intense in colour and cover large areas of the leaves.

Source of infection and spread

Aphids spread the virus from plant to plant, needing only very short feeding times for transmission. Many aphid

Fig 2.40 Alfalfa mosaic on lettuce. Note the large areas of bleached tissue.

Fig 2.41 Alfalfa mosaic on capsicums showing yellow, blotchy leaves.

Fig 2.42 Calico disease caused by *Alfalfa mosaic virus.*

species are able to transmit the virus, including the green peach aphid (*Myzus persicae*), the pea aphid (*Acyrthosiphon pisum*), the bluegreen aphid (*Acyrthosiphon kondoi*) and the spotted alfalfa aphid (*Therioaphis trifolii*). Annual crops are most likely to become infected when planted near perennial hosts such as lucerne and clover that often have a high level of infection.

Management

The level of alfalfa mosaic infection in vegetable crops is usually not sufficient to cause serious losses. Adequate control is possible by not planting susceptible crops adjacent to stands of lucerne and clover.

■ *TOMATO SPOTTED WILT VIRUS* AND RELATED VIRUSES

Tomato spotted wilt virus (TSWV), first described from tomato in Australia in 1915, is the type or reference virus for the Tospovirus group of plant viruses. Tospoviruses cause major economic losses in a wide range of vegetable, field and ornamental crops throughout the world. At least 16 tospoviruses are known worldwide and all are transmitted by thrips. An important reason for the increased importance of the group has been the distribution worldwide of the western flower thrips (*Frankliniella occidentalis*), a very efficient vector or carrier of TSWV and several other tospoviruses. Three tospoviruses have been identified in Australia: TSWV, *Capsicum chlorosis virus* and *Iris yellow spot virus.*

Tomato spotted wilt virus (TSWV)

TSWV occurs throughout Australia. It has one of the largest host ranges of any plant virus, infecting over

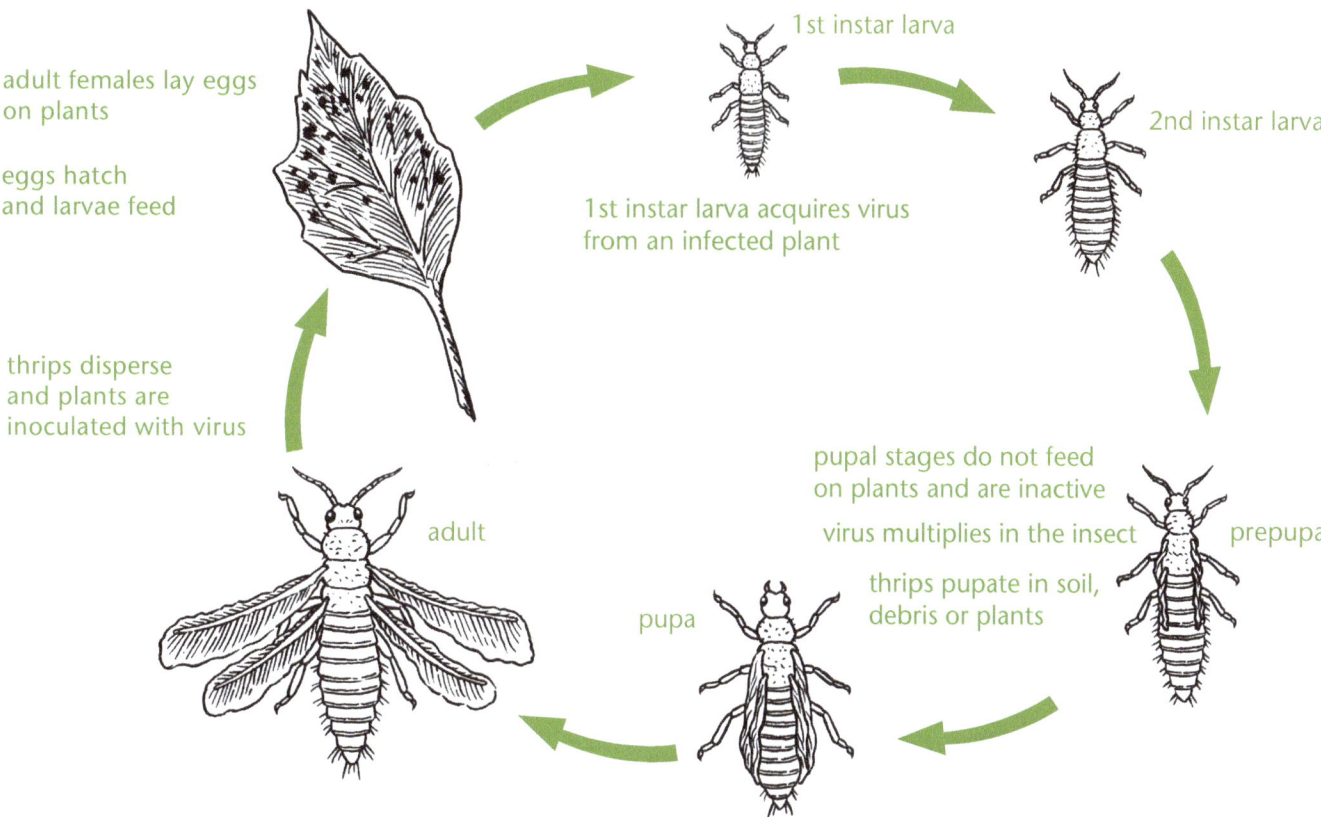

Fig 2.43 Life cycle and virus transmission of *Tomato spotted wilt virus* (TSWV) by thrips.

900 species of weeds, field crops, vegetables and ornamentals. Many hosts are in the potato (Solanaceae), aster (Asteraceae) and legume (Fabaceae) families.

Crops that frequently suffer major losses from TSWV in Australia include lettuce, capsicum, tomato, potato, and ornamental crops such as aster, statice, calendula and dahlia. Other crop hosts that have sporadic losses include peanut, tobacco, chickpea, rhubarb, eggplant, celery and a range of ornamental species. Weeds have a major role in the life cycle of both the virus and the thrips vectors. Many weed hosts of TSWV occur in Australia and include sowthistle (*Sonchus oleraceus*), capeweed (*Arctotheca calendula*), thornapples (*Datura* spp.), cobbler's pegs (*Bidens pilosa*), nightshades (*Solanum* spp.) and Jamaican snakeweed (*Stachytarpheta jamaicensis*).

TSWV causes a range of symptoms, depending on the plant species affected. Symptoms are influenced by the age of plants, weather conditions and nutritional status. Symptoms include ringspots, mottling, chlorotic blotches and line patterns on leaves. Both leaves and fruit are often distorted with dark spots or ring patterns on fruit. Wilting and purpling of leaves can occur and necrotic lesions can develop on stems of affected plants.

Capsicum chlorosis virus (CaCV)

Capsicum chlorosis virus is a member of the *Watermelon silver mottle virus* or serogroup IV group of Tospoviruses. Members of this group are widespread and damaging throughout Asia. CaCV was first described from Queensland in 1999 and occurs in all coastal and sub-coastal vegetable production areas in Queensland. The virus has also been found in north-eastern, coastal areas of NSW and at Kununurra in the Ord River area of Western Australia.

In Australia, CaCV infects capsicum, chilli, tomato, Hoya and peanut. It also infects several weed species including

Fig 2.44 Typical ringspot symptoms of TSWV on tomato leaves.

Fig 2.45 A lettuce crop abandoned due to TSWV infection.

Fig 2.46 Symptoms of TSWV on a range of capsicum fruit.

Brazil, Israel, Japan and Europe and was first found in Australia in 2003.

On onion, IYSV causes eye-like or diamond-shaped spots on leaves and the seed stalk, which often bends as the spot dries causing the seed stalk to collapse.

Transmission and spread of tospoviruses

All viruses in the Tospovirus group are transmitted by thrips, minute, slender insects belonging to the insect Order Tysanoptera. The thrips/tospovirus relationship is very specific and of the many thousands of species of thrips identified throughout the world, less than 20 species are known to transmit tospoviruses. The four vector (or carrier) species found in Australia are:

- western flower thrips (*Frankliniella occidentalis*).
- tomato thrips (*Frankliniella schultzei*).
- melon thrips (*Thrips palmi*).
- onion thrips (*Thrips tabaci*).

Fig 2.47 TSWV symptoms on snake weed (right), and cobbler's pegs, common weed hosts of the virus in Queensland.

Ageratum conyzoides (Billygoat weed), which is a common and symptomless host throughout coastal Queensland.

CaCV is transmitted by melon thrips and tomato thrips in Australia and in Thailand by *Ceratothripoides claratris*, a thrips species not found in Australia.

The symptoms of CaCV are somewhat similar to those of TSWV but distinct differences do occur. Younger leaves develop marginal and interveinal chlorosis and are narrowed and curled often giving leaves a strap-like appearance. Older leaves are chlorotic and frequently develop ringspot and line patterns typically seen with TSWV. Fruit from infected plants is small, distorted and frequently has necrotic spots and scarring over the surface. Infected plants are stunted and do not develop the chlorotic blotching generally seen on young leaves of plants infected by TSWV.

Iris yellow spot virus (IYSV)

IYSV is primarily a pathogen of onion and related *Allium* species and transmission appears to be exclusively by the onion thrips (*Thrips tabaci*). The virus occurs in the USA,

Tospoviruses are not spread by other sap-sucking insects such as aphids, whiteflies and leafhoppers.

Transmission of tospoviruses can only occur if they are acquired from infected plants by first or early second instar larvae thrips. More mature thrips, including adults, may acquire the viruses but the viruses cannot complete their life cycle within the insect to allow transmission. The larvae can acquire the virus during feeding periods of less than 30 minutes.

Once acquired by immature thrips, the viruses circulate and multiply within the insect and are transmitted to plants as the adult thrips pierce and suck the contents of plant cells. Thrips remain infective for life but do not pass the virus to their offspring through the egg.

About five days are required from the time the virus is acquired from an infected plant until the thrips is able to transmit the virus to another plant. This allows time for the virus to move and multiply in the insect gut and salivary glands. Long feeding periods are not required for thrips to transmit the virus and efficient transmission occurrs in five to 10 minute feeding periods.

Tospoviruses are not spread in seed or on cutting and cultivation equipment. They are not spread by handling plants and do not survive in soil or decaying crop residues. Tospoviruses can be spread in infected plant parts used for propagation such as cuttings and bulbs.

Management of tospoviruses

Infected plants cannot be cured. Management aims to prevent or reduce the levels of disease in crops by removing or avoiding sources of virus infection and minimising spread by thrips.

Crop and farm hygiene. Old, infected crops infested by thrips are a major source of virus and should be removed as soon as possible, particularly if young crops are to be planted nearby. Weeds along headlands and irrigation channels provide host plants for thrips and tospoviruses. Disease levels are often higher adjacent to these areas. Maintaining a buffer zone of at least 25 m, which is free of weeds between a virus source and a susceptible crop can considerably reduce virus levels.

Controlling thrips with insecticide. Reducing thrips populations with appropriate insecticides can help reduce virus spread. However, insecticides are often of limited value in tospovirus control because there is significant spread from outside sources and only a short feeding time is required for transmission. Frequent use of insecticides may also lead to thrips populations developing resistance.

Use healthy planting material. Viruses can be introduced in infected seedling plants which then provide a virus source through the life of the crop. Seedling production areas should be located well away from production areas, kept weed-free and a regular insect management system used. Thrips-proof netting or UV-absorbing plastic provides a higher level of protection for seedling production.

Resistant varieties. Varieties of capsicum and tomato resistant to TSWV are available. If using these varieties, care should still be taken with crop–farm hygiene and other preventative measures to prolong the useful life of the resistance sources.

■ FURTHER INFORMATION

Persley DM (2007). *Thrips and tospoviruses – a management guide.* Department of Primary Industries and Fisheries, Brisbane.

Persley DM, Thomas JE and Sharman M (2006). *Tospoviruses – an Australian perspective. Australasian Plant Pathology* **35**, 161–180.

Asian vegetable production is a growing and dynamic sector within the Australian vegetable industry. The diverse lines grown are grouped into leafy and non-leafy types. The leafy Asian vegetables are predominantly from the genus *Brassica*. Many common names apply to these vegetables, but they can be grouped as pak choy (*Brassica rapa*), bok choy (*B. rapa*), choy sum (*B. rapa*), Chinese cabbage (*B. rapa*), gai choy (Asian mustard greens, *B. juneca*) and gai lan (Chinese broccoli or Chinese kale, *B. oleracea*).

Non-leafy types include snake bean and various members of the cucurbit family, including luffa, bitter melon, gourds and horned cucumber.

Disease prevention and control is a critical aspect of leafy Asian vegetable production as the crops are susceptible to a range of bacterial, fungal and viral diseases. In the absence of good crop protection strategies, these diseases can have a serious effect on production.

This chapter covers the more important diseases affecting leafy Asian vegetables, snake bean, okra and taro. The non-leafy cucurbit types are susceptible to the same range of diseases as the more widely grown cucurbits such as melons, pumpkins and squash. Readers are referred to the Chapter on cucurbit diseases for this information.

LEAFY ASIAN VEGETABLES

■ BACTERIAL LEAF SPOTS AND BLACK ROT

Cause
The bacteria *Xanthomonas campestris* and *Pseudomonas syringae* pv. *maculicola*.

Symptoms
The first symptoms are usually small, yellow V-shaped areas developing along the leaf margin. These areas soon turn brown and dry out. Vein blackening may extend down the leaves into the petiole and the stem. When the stem is cut across, a black ring will be seen in the water-conducting tissues just beneath the bark.

The V-shaped marginal lesions develop when the bacteria enter the leaf in water through hydathodes, natural openings at the vein-ends on leaf margins. When bacteria gain direct entry to the vascular system through wounds, root injury or hail damage, yellowing of the leaf blade and leaf collapse may occur before the leaf margin symptoms develop.

Black rot is caused by the bacterium invading fleshy leaves and petioles. Soft-rotting organisms often quickly follow and then the plants rot rapidly.

Certain strains of the bacterium infect plants through the leaf stomata (breathing pores) causing scald symptoms. Tan, circular spots with yellow haloes develop between the veins. These coalesce and dry out to give the leaf scald symptoms. Affected areas often crack and disintegrate, giving the leaf a tattered appearance.

Source of infection and spread
Usually, the bacterium is introduced in seed but may survive in undecomposed crop residues, brassica weeds and

Fig 3.1 Bacterial leaf spots on choy sum.

Fig 3.2 Bacterial leaf spot on Chinese cabbage.

other brassica crops. The bacteria spread in water splash during wet, windy weather or by overhead irrigation. The pathogen can also disperse with insects and on people and equipment moving through a crop. Warm, wet weather favours disease development.

Importance
Black rot is one of the most serious diseases of brassicas worldwide, particularly in warm, humid climates.

Management
- Use seed that treated in hot water to reduce seed-borne infection.
- Keep seedling production areas separate from field production areas.
- Reduce sowing rates to prevent overcrowding the seedlings.
- Keep seedling production areas free from susceptible weeds.

- Plough in diseased crops immediately after harvesting. Rotate crops so that brassicas are grown only once in every three crops in each field.

- Control weeds and insects.

- Apply a regular, preventative spray program of copper to limit spread of black rot, particularly in seedlings. For best results, copper should be tank-mixed with mancozeb.

■ BACTERIAL SOFT ROT

Cause
Several species of bacteria including *Erwinia* spp. and *Pseudomonas* spp.

Symptoms
A slimy soft rot develops on the heads of Chinese cabbage. An unpleasant odour is often associated with affected heads.

Source of infection and spread
The bacteria occur with decaying organic matter in the soil. Bacteria invade Chinese cabbage heads, often through injury sites caused by weather damage, insects, tools or with harvesting. Hot weather favours the development of rot, which may be extremely severe if wet weather occurs. In the field, spread is by water splash and on cutting knives. The bacteria can spread by contact during transit and storage. Severe head rot appears to be associated with high application rates of nitrogen fertiliser, particularly chicken manures.

Importance
Head rot is often a major cause of loss in Chinese cabbage, particularly if wet weather occurs during the late summer production period.

Fig 3.3 Bacterial leaf spot.

Fig 3.4 Symptoms of bacterial soft rot on Chinese cabbage in the field.

Fig 3.5 Bacterial soft rot on cut stems and showing internal symptoms.

Fig 3.6 Phoma leaf spot, showing pycnidia.

Management

- Avoid harvesting when crops are wet.

- Avoid high rates of nitrogen fertiliser application.

- Sterilise cutting knives if the disease occurs in the field.

- Remove field heat from heads as quickly as possible after harvest and store at 0°C, 95% relative humidity.

- Susceptibility to head rot differs between varieties of Chinese cabbage.

■ ALTERNARIA SPOT

Refer to the description of alternaria spot in the chapter on brassicas.

■ PHOMA LEAF SPOT AND STEM CANKER (BLACK LEG)

Cause
The fungus *Phoma lingam* (*Leptosphaeria maculans*).

Symptoms
Plants can be infected at any stage of growth. Symptoms may occur on cotyledons or the first true leaves of seedlings, resulting in damping-off if the infection is severe.

On leaves, symptoms can resemble alternaria leaf spot, however when the lesions are viewed with a hand lens, *Phoma* spots have tiny, black, dome-shaped fungal fruiting bodies or pycnidia. The pycnidia can exude a pink spore mass under humid conditions.

On stems, large, brown to black, sunken cankers occur near the base of the stem, eventually girdling it and causing

Fig 3.7 Phoma stem canker symptoms.

stunting and the plants wilt. Pycnidia of the fungus may appear on the affected area. Internally, the stem shows a brown, dry rot.

Source of infection and spread
The fungus is commonly introduced in infected seed and also survive on undecomposed crop residues. Low levels of seed infection (<1%) are sufficient to cause serious disease outbreaks. Wet leaves are required for infection. Large numbers of spores, produced in pycnidia, spread in rain or irrigation. The fungus may also spread by air borne ascospores produced on infested crop residues. Temperatures of 15–20°C are optimal for ascospore infection.

Importance
Black leg is an important disease of vegetable brassica crops worldwide. It is more common and damaging in cooler climates.

Management
- Use seed treated in hot water (50°C for 25 minutes).

- Produce seedling transplants away from production fields and maintain high standards of nursery practice and hygiene.

- Plough in diseased crops immediately after harvest and ensure that all residues have thoroughly decomposed before replanting with brassicas.

- Rotate crops so that brassicas are not grown on the same land more than once every three to four years.

■ CLUB ROOT

Refer to the description of club root in the chapter on brassicas.

Fig 3.8 Club root on Chinese cabbage in field.

Fig 3.9 Club root on root system.

Fig 3.10 Downy mildew.

■ DOWNY MILDEW

Refer to the description of downy mildew in the chapter on brassicas.

■ WHITE LEAF SPOT

Cause
The fungus *Pseudocercosporella capsellae*.

Symptoms
Many small (1 cm wide), yellow spots develop on leaves. The spots have a whitish, papery appearance as the leaves age. Leaf spots are often surrounded by a dark brown margin.

Fig 3.11 Aggregations of downy mildew spores.

Fig 3.13 White leaf spot symptoms.

Source of infection and spread

The fungus survives from season to season on undecomposed crop residues in the soil. Spores produced in the pycnidia eject forcibly and disperse in the wind. Cool, wet weather favours infection and disease development. The fungus is also seed-borne.

Importance

White leaf spot is one of the most significant diseases of leafy Asian vegetables in winter. Leaf spotting reduces produce marketability, and, if the disease occurs on the seedling stage, it can cause widespread plant death.

Management

- Plough in diseased crops immediately after harvesting. Do not replant until all residues have completely decomposed.

- Rotate brassica crops with unrelated species.

- Apply the recommended fungicides. Base spray programs on regular applications of protectant fungicides, with targeted applications of systemic products following cool, moist weather.

■ WHITE BLISTER (WHITE RUST)

Refer to the description of white blister in the chapter on brassicas.

Importance

White blister occurs almost wherever brassica crops are grown. Brassica crops affected include Chinese cabbage, pak choy, bok choy, choy sum, gai choy and gai lan.

The presence of blisters on the stem and the surface of the leaf can reduce product quality and can render product as unmarketable.

Fig 3.12 White leaf spot on bok choy.

Fig 3.14 White blister.

Fig 3.15 White blister symptoms on Chinese cabbage.

The pathogen occurs as a complex of races distinguished by their ability to infect various brassica species. For example, the races infecting several brassica weed species differ from those infecting broccoli and cauliflower, while races from radish are unlikely to infect other brassica species.

Management
- Use high quality seed and plant disease-free transplants.
- Ensure that all crop residues decompose thoroughly before sowing.
- Rotate crops. Avoid consecutive plantings of brassicas and rotate with non-brassica crops.
- Avoid long periods of leaf wetness as spores need to water to germinate.
- Maintain good air movement within crops through row orientation, increased space between plants and avoiding areas prone to heavy fog and mist.

- Apply recommended fungicides. Apply these before the disease becomes serious and use as part of a fungicide-resistance management strategy. White blister has the capacity to develop resistance to fungicides very quickly, particularly where systemic fungicides are over-used and/or not combined, or alternated with, protectant chemicals.

■ DAMPING-OFF AND FUNGAL ROOT ROTS

Refer to the chapter on Common diseases of vegetable crops.

■ TURNIP MOSAIC VIRUS

Cause
Turnip mosaic virus (Potyvirus).

Fig 3.16 *Turnip mosaic virus* in bok choy.

Symptoms

Yellow ringspots and mosaics develop on leaves. The mosaic markings can develop into yellow or brownish spots surrounded by irregular necrotic rings on older leaves. Symptom development is highly dependent on environmental conditions; in cooler temperatures, the markings become distinctly necrotic.

Source of infection and spread

The virus has a wide host range among cultivated brassica and brassica weed species. It also infects crop and weed hosts in other plant families.

The virus is spread by aphids and is acquired and transmitted during short feeding probes.

Importance

Mosaic virus can cause severe losses in Asian leafy brassicas, especially in the cooler winter months.

Management

- Control weeds around the crop area.

- Plant healthy transplants.

- Control aphids in the crop as required.

- Remove diseased plants showing signs of the disease to prevent the virus spreading to the rest of the crop.

OKRA

Okra (*Abelmoschus esculentus*) is in the family Malvaceae.

■ CERCOSPORA LEAF SPOT

Cause

The fungus *Cercospora* sp.

Symptoms

A sooty to dark olivaceous mould develops on the lower leaf surface. As the disease progresses, leaves roll, wilt and fall from the plant.

Source of infection and spread

The fungal spores are air-borne with disease development favoured by warm, humid weather.

Importance

Cercospora leaf spot is a common and often serious disease during prolonged wet, humid weather.

Management

- Apply recommended fungicides.

■ POWDERY MILDEW

Cause

The fungus *Podosphaera xanthii*.

Symptoms

White powdery spots appear mainly on the upper surfaces of older leaves and plant stems. The lesions can coalesce to cover the entire leaf surfaces. Heavy disease pressure causes leaf chlorosis, necrosis and defoliation.

Source of infection and spread

The fungus requires a living host for survival and can infect numerous weed and crop species. The white leaf spots comprise masses of fungal spores blown long distances on the wind to infect susceptible plants. Infection is favoured by periods of heavy dew or high humidity, but the disease can also become severe under dry conditions.

Importance

The disease has the potential to be severe, particularly under dry conditions towards the end of the growing season.

Management

- Apply regular applications (every seven to 10 days) of recommended protectant fungicides to minimise infection and disease spread. For best results, apply sprays prior to symptom development.

- Destroy old crops promptly to prevent disease carry-over to younger crops.

■ VERTICILLIUM WILT

Cause

The fungus *Verticillium dahliae*.

Symptoms

The lower leaves turn pale or are blotchy with pale green or yellow areas. As the disease progresses, the leaf margins and tips turn yellow and chlorotic, angular, interveinal, V-shaped lesions develop. In time, the leaf margins and yellow areas on leaves become necrotic. Infected plants are stunted and unthrifty. If the lower stems are cut longitudinally, a brown discolouration

Fig 3.17 Verticillium wilt in okra.

of the water-conducting tissues is evident. The discolouration may also extend into the petioles of the lower leaves. Disease symptoms may be accentuated if the plant is under stress due to fruit load or some other factor.

Source of infection and spread
The fungus is a common soil inhabitant and survives in the soil for long periods as microsclerotia. It enters the roots, usually through wounds, and moves into the water-conducting tissues of the stem. Cool weather favours the disease and the optimal temperature range is 20–24°C. *Verticillium* infects other crop plants and weeds including tomato, potato, cotton and peanut. Susceptible weeds include Noogoora burr (*Xanthium pungens*), stinking roger (*Tagetes minuta*) and cobbler's pegs (*Bidens pilosa*).

Importance
Two races of the fungus occur in Queensland. Verticillium wilt is rarely seen in northern production areas.

Management
• Avoid growing okra in rotation with other susceptible crops such as potatoes and tomatoes.

• Keep crop areas free from volunteer plants, Noogoora burr and other susceptible weeds.

• Take steps to avoid moving the fungus from infested to clean areas on vehicles and machinery.

■ ROOT-KNOT NEMATODE

Refer to the chapter on Common diseases of vegetable crops.

SNAKE BEAN

Snake bean (*Vigna unguiculata* ssp. *sesquipedalis*), also known as long bean or yard-long bean, is a member of the legume or Fabaceae family. Grown throughout the tropics and subtropics, it is one of the most important vegetable crops in Asia. In Australia, this tall, climbing annual is grown mainly around Darwin in the Northern Territory and coastal, northern Queensland.

The most serious disease problems in snake bean in Australia are damage from root-knot nematodes and Fusarium wilt disease. Powdery mildew, cercospora leaf spot and rust are common and sometimes damaging in crops.

■ FUSARIUM WILT

Cause
The fungus *Fusarium oxysporum* f.sp. *tracheiphilum*.

Symptoms
Symptoms usually appear at flowering, and the start of pod set. Leaves on infected plants turn yellow and fall. The

Fig 3.18 Fusarium wilt symptoms on snake bean.

Fig 3.19 Fusarium wilt symptoms on a section of snake bean stem.

plant wilts over several days and then dies. A characteristic symptom of Fusarium wilt is the reddish-brown discolouration of the water conducting tissue of the stem and roots, seen when these parts are cut with a sharp knife.

Source of infection and spread

The fungus can spread in contaminated soil and in infected seed. Once introduced, the pathogen can survive in the soil for decades, even in the absence of susceptible crops. The fungus infects through the roots, particularly where damage has occurred from cultivation or root-knot nematodes.

Importance

Fusarium is a major cause of crop failure of snake beans in the Darwin area.

Management

- Avoid introducing the pathogen into production areas on contaminated seed, implements, footware or in contaminated water.

- Although commercially acceptable, resistant varieties are not available, effective management is possible by grafting onto Fusarium-resistant snake bean rootstocks, for example the variety 'Iron'. This variety is also resistant to root-knot nematode. When planting grafted plants it is essential to keep the graft union well above the soil level and keep all parts of the plants secured on trellises or stakes to prevent soil contact, which can lead to Fusarium infection.

■ ROOT-KNOT NEMATODE

Refer to the chapter on Common diseases of vegetable crops.

TARO

Taro (*Colocasia esculenta*) is in the Araceae family.

PHYTOPHTHORA LEAF BLIGHT – BIOSECURITY THREAT

Cause

The oomycete *Phytophthora colocasiae*.

Symptoms

Small (1–2 cm diameter), circular, purple to brown, water-soaked lesions develop on the leaves. The lesions acquire a zonate appearance and rapidly expand until the entire leaf is affected. A white ring of spores called sporangia, form around the edges of the lesions. In severe infections, the pathogen moves from the leaf lamina to invade the petiole and corm, where it causes a firm, brown rot with a sharply defined margin.

Source of infection and spread

Phytophthora colocasiae is favoured by warm weather. Temperatures ranging between 27–30°C are optimal for infection. The sporangia spread from the edges of the leaf lesions in rain-splash and wind-blown rain. They germinate on leaves or petioles or wash into the soil where they can infect corms directly. Motile zoospores

Fig 3.20 Phytophthora leaf blight on taro.

Fig 3.21 Close view of Phytophthora leaf blight on a taro leaf.

are released from the germinated sporangia and swim through wet soils to infect susceptible taro plants. Infected planting material, along with sporangia and zoospores in wind-driven rain, are the main sources of inoculum in new plantings.

Importance

This pathogen is both widespread and highly destructive in the tropical regions of South-East Asia, the Pacific, Africa, the Caribbean and the Americas. The warm, humid days and cooler wet nights in the tropics are ideal for the proliferation and spread of this pathogen. Epidemics of leaf blight may occur throughout the year during rainy, overcast weather. When conditions are optimal for disease, entire crops may become blighted within seven days.

What to do if you suspect Phytopthora leaf blight

This pathogen is a biosecurity risk to Australia. Any suspected affected plants should be reported to the nearest Department of Primary Industries or the Plant Health Australia hotline (1800 084 881).

■ FURTHER READING

Brooks F (2005). *Plant disease lessons – taro leaf blight. American* Phytopthological Society, St. Paul, Minnesota. http://www.apsnet.org/education/lessonsPlantPath/taroleafblight/default.htm

Ooka JJ (1994). *Taro diseases: A guide for field identification.* Research Extension Series 148, Hawaii Institute of Tropical Agriculture and Human Resources, Hawaii.

Asparagus (*Asparagus officinalis*) is a perennial monocot in the family Asparagaceae. The crop is grown for its immature, unexpanded shoots called spears, which originate from underground crowns (rhizomes). Spears are usually harvested in spring, and the remaining spears are allowed to mature and develop into the foliage, or fern, of the plant. The fern provides energy for the crown, allowing the plant to grow and healthy plants produce economic yields for up to 20 years. The crop is usually established by planting one-year-old crowns previously established from seed in a nursery bed.

FUNGI

■ ANTHRACNOSE

Cause
The fungus *Colletotrichum gloeosporioides*.

Symptoms
Dark, water-soaked, oval-shaped lesions develop on the fern and form concentric raised rings as the lesions age. Orange-pink fungal spores are visible in the lesions under wet, humid conditions. The lesions can girdle the stem, killing the fern.

Source of infection and spread
The fungus survives on crop debris and has a wide range of crop and weed hosts. Spores spread in water splash

Fig 4.1 Anthracnose lesions on spears. Note wet rot, pale fungal mycelium and the darker, fungal spores.

Fig 4.2 Mature anthracnose lesion on asparagus canes, showing dark fungal fruiting bodies.

and disease development is favoured by wet, humid conditions.

Importance
Anthracnose has been found only on asparagus grown under hot, humid conditions in northern Australia where it can cause crop failure.

■ CERCOSPORA BLIGHT

Cause
The fungus *Cercospora asparagi*.

Symptoms
Symptoms develop first on the lower fern as oval-shaped spots with light grey centres and a red-brown to purple margin. As the spots age, they become grey-white. The disease progresses up the fern causing defoliation and girdling.

Source of infection and spread
Fungal spores are dispersed by wind and water splash. Warm, wet weather favours disease development. The fungus survives on crop residues and debris.

Importance
Cercospora blight can be a serious disease in hot, humid production areas.

Fig 4.3 Stem rot caused by *Fusarium moniliforme*.

■ FUSARIUM CROWN, ROOT AND STEM ROT

Cause
The fungi *Fusarium* spp.

Symptoms
Recently planted crowns fail to establish, ferns yellow and crowns rot causing plant death. Sections of fern become parched and brown. At the base of affected ferns, the stem usually cracks longitudinally and has a dark-red discolouration. Internally, stem tissues are rotted and show dark, reddish-brown discolouration. Similar discoloured areas occur on roots, which may also develop cracking.

Source of infection and spread
The fungi are distributed widely and survive in soil and crop debris for many years. The fungi infect many crops, including sorghum, maize, rice, peanut and rockmelon. Spores are produced on infected plant material and are transported by wind and water. The fungi can occur on asparagus seed coats and in new nursery crowns before transplant.

Importance
Fusarium crown, root and stem rot occur in all production areas.

■ PHOMOPSIS STEM BLIGHT

Cause
The fungus *Phomopsis asparagi*.

Symptoms
The first symptom of stem blight is discolouration of the stem tissue, followed by the appearance of oval-shaped

Fig 4.4 Phomopsis stem blight showing stem lesions and fern death.

lesions with light brown centres and slightly darker margins. As the disease progresses, affected areas shrivel and become well-defined lesions surrounded by dark margins. Lesion centres eventually becomes ashy-white, containing masses of spores that are released and infect new growth.

The disease affects the leaves and stem, which may be girdled and die if stem bases are affected.

Source of infection and spread
The fungus survives on plant debris, for at least six months and on volunteer asparagus plants. Spores spread during wet, windy weather and disease development is favoured by warm, wet conditions.

Importance
Stem blight is a serious disease of asparagus in wet, humid production areas. It can cause defoliation in the fern and plant death.

Fig 4.5 Phomopsis lesion on an asparagus spear.

Fig 4.6 Phomopsis lesions on asparagus stems.

■ PHYTOPHTHORA SPEAR AND CROWN ROT

Cause
The oomycete *Phytophthora megasperma*.

Symptoms
The initial symptoms are soft, water-soaked lesions or spots at or near the soil line. Lesions elongate until the spear curves and collapses. Spear tissue becomes soft and may be covered by white fungal mycelium. Infected roots may turn brown and rot. The pathogen can also infect crowns, where it causes browning and rotting.

Source of infection and spread
The pathogen is soil borne and survives for long periods in the soil as resistant oospores. Zoospores are released from germinating oospores or from sporangia on infected host tissue and infect shoots and roots. Young spears are very susceptible. The pathogen is favoured by cool, wet soils.

Importance
The disease is most likely to be serious in young spears produced during cool, wet weather and in crops grown in poorly drained, heavy soils.

■ PURPLE SPOT

Cause
The fungus *Stemphylium vesicarium*.

Fig 4.7 Purple spot caused by the fungus *Stemphylium*.

Fig 4.8 Purple spot lesions on asparagus spears.

Fig 4.9 Rhizoctonia symptoms on spears.

Symptoms

Small, purple spots occur on ferns and occasionally on spears. On ferns, the spots enlarge to light brown lesions with purple margins. The lesions may completely girdle the stems, resulting in the death of sections of fern.

Source of infection and spread

The fungus survives on ferns and crop debris and produces ascospores from small black fruiting bodies (perithecia) in spring, which infect the spears causing purple spots. Disease development continues under warm, moist conditions. At least 12 hours of surface wetness is required at temperatures between 10–20°C for infection to occur. High relative humidity also favours disease development. New infections occur when spores disperse in wind and water splash.

The fungus can also survive in seed and on volunteer asparagus plants.

Importance

Purple spot is a common disease in asparagus and is most severe during persistent, showery weather. The presence of purple spots on spears makes them unmarketable.

■ RHIZOCTONIA CROWN ROT

Cause

The fungus *Rhizoctonia solani*.

Symptoms

A dark brown rot occurs at the base of the plant, extending out along the roots. Affected spears show small, red lesions and kinked growth.

Source of infection and spread

Rhizoctonia solani is a common soil inhabitant. In asparagus, the fungus most commonly enters the plant by colonising wounds made when spears are harvested. It grows down to the base of the plant, potentially causing general decay. Spears are infected as they grow through roots or stems affected by the fungus. Base rot is favoured by sandy soils, warm temperatures and wet conditions.

Importance

Rhizoctonia crown rot is a relatively common disease in asparagus. Control measures are not available.

Fig 4.10 Base and root rot caused by *Rhizoctonia*.

Fig 4.11 *Rhizoctonia* causing kinked growth in spears.

■ RUST

Cause
The fungus *Puccinia asparagi*.

Symptoms
The pathogen has several spore stages during its life cycle and these produce different symptoms on plants. The orange spore stage occurs in spring and causes light green patches on new spears, maturing to yellow or pale orange pustules in concentric rings. In late spring and early summer, when conditions are warm and moist, spores produced during the orange spore stage infect new fern growth and develop brick-red pustules associated with the red spore stage on fern stalks, branches and leaves. These red pustules produce rust-red coloured spores

Fig 4.12 Urediniospores of the rust fungus *Puccinia asparagi* on asparagus canes.

Fig 4.13 Teliospores of the asparagus rust fungus on canes.

(urediniospores) in a powdery mass that can reinfect the fern and increase disease incidence. Ferns will then turn yellow and brown, causing defoliation and dieback. The black spore stage may develop as ferns mature and senesce or when autumn weather begins. The same pustules that produced the red spores begin producing the black spores. The pustule's appearance will slowly change to a jet-black powdery mass associated with the black spore (teliospore) over-wintering stage of the fungus. These spores provide the source for new rust fungus infections in the next season.

Source of infection and spread
Fungal spores disperse in wind, often over considerable distances. Warm temperatures and moisture, including morning dews, high humidity and rain, favour disease development. Dry, hot weather inhibits the disease.

This species of rust is confined to edible and ornamental asparagus species.

Importance
Rust occurs in most asparagus production areas worldwide.

The disease causes defoliation, poor vigour and loss of production.

■ DISEASE MANAGEMENT IN ASPARAGUS CROPS

- Disinfect seed before planting.
- Plant only healthy crowns.
- Plant crops in well-drained and well-structured soils.

- Reduce humidity in plantings by using wider row widths and orientate rows to allow wind movement that will dry ferns and reduce humidity.

- Maintain good growth and crop health with adequate irrigation and fertiliser applications. Irrigate with trickle systems at the base of plants rather than overhead.

- Crop residues and plant debris are the main source of inoculum for most asparagus diseases. Remove crop residues and debris between seasons. Destroy volunteer asparagus plants in fallow land and around production areas. Ensure crop residues are completely broken down before planting new crops.

- Apply appropriate fungicides for the control of rust, purple spot and anthracnose. Monitor crops and apply chemicals when diseases are detected, to maximise control.

■ FURTHER INFORMATION

Irvine G, Malipatil M, Edwards J, White D, Murdoch C and Harrison G (2004). *Pests, diseases, weeds and disorders of asparagus in southern Victoria*. State Government of Victoria, Department of Primary Industries, Knoxfield, Victoria.

Common or French bean (*Phaseolus vulgaris*) belongs to the family Fabaceae (Leguminosae) and is a major food legume throughout the world. Common beans are grown for green pods, which are sold in fresh markets or used for freezing and canning.

Beans are warm season annuals and grown during summer in temperate areas and during the cool dry season in the tropics and subtropics.

Beans are susceptible to a range of bacterial, fungal and viral pathogens. High quality seed, crop rotation and farm hygiene are important aspects of profitable bean production.

BACTERIA

■ BACTERIAL BROWN SPOT

Cause
The bacterium *Pseudomonas syringae* pv. *syringae*.

Symptoms
Small (3–8 mm diameter) tan spots with reddish-brown margins develop on leaves. The spots are usually circular but may be angular when bordered by leaf veins. Spots are often surrounded by a bright yellow zone. The centre of the spot may eventually fall out, giving the leaves a ragged appearance. Tan spots with reddish-brown margins may occur on stems.

Lesions on pods are initially circular, dark green and water-soaked. They then enlarge and become sunken and tan with distinctive, reddish-brown margins. If the pods are infected early, they may become twisted or bent where lesions develop.

Source of infection and spread
Sowing contaminated seed is the most important means of introducing the pathogen. Seed contamination arises by direct pod infection or contact of the seed coat with bacteria in plant debris during harvesting and handling.

The bacteria can also survive and multiply on healthy plants for a long time as epiphytic populations that do not cause disease symptoms. When environmental conditions that favour infection occur, epidemics may quickly develop. Cool, wet, windy weather favours brown spot disease. Periods of intense rainfall are particularly conducive to disease development. Once the disease is established, wind, rain, overhead sprinkler irrigation and the movement of workers and machines through the wet crop assist disease spread. The bacterium often enters leaves through rust pustules and hail or wind-damaged areas. It survives on undecomposed crop residues and possibly other host plants.

Importance
Bacterial brown spot occurs sporadically and usually is serious only in damaged crops.

Management
• Use disease-free seed, such as certified or approved bean seed.

Fig 5.1 Bacterial brown spot on bean leaves.

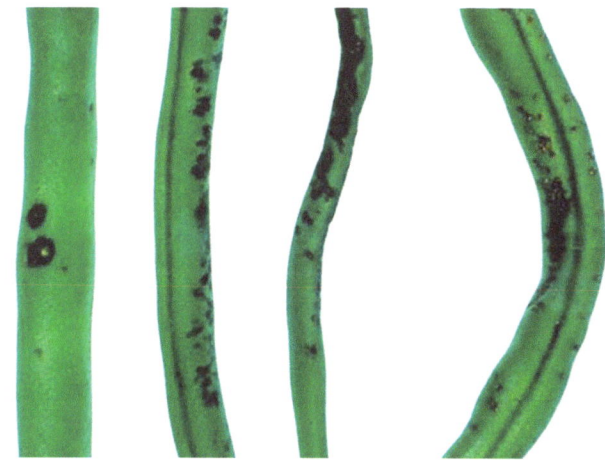
Fig 5.2 Bacterial brown spot on bean pods.

- Avoid moving workers and machinery between diseased and disease-free areas of the crop, particularly in wet weather.

- Use a regular protectant copper spray program to slow disease development. Copper should be tank-mixed with mancozeb for best results.

- Plough in diseased crops immediately after harvesting.

- Avoid very susceptible varieties.

■ COMMON BACTERIAL BLIGHT

Cause
The bacterium *Xanthomonas campestris* pv. *phaseoli*.

Symptoms
Leaf symptoms initially appear as small, angular, water-soaked spots, which gradually enlarge to form large, brown, dead areas on the leaves. These areas are often surrounded by a bright yellow zone. Spots develop on the margin and in interveinal areas of the leaf. As spots enlarge and coalesce, the plants appear burnt.

Dark green, water-soaked streaks may develop on stems, and later become tan-coloured.

Pod symptoms consist of small, water-soaked spots later becoming sunken and reddish-brown. Under humid conditions, pod lesions often become covered with yellowish bacterial ooze. The symptoms on pods may be difficult to distinguish from those of halo blight.

Source of infection and spread
Contaminated seed is the most important source of the pathogen. Seed contamination arises by direct pod infection or contact of the seed coat with bacteria in plant debris during harvesting and handling. Both susceptible and tolerant varieties can harbour the pathogen.

Warm (28–32°C), humid, showery or wet weather favours infection. Wind-blown rain, soil and plant debris will spread the bacteria, as will contact between wet leaves, overhead irrigation, and the movement of people and machinery through crops.

The bacterium can survive on undecomposed crop residues, particularly residue at or near the soil surface. It can also survive on other hosts such as navy bean, mung bean (*Vigna radiata*), phasey bean (*Macroptilium lathyroides*) and lablab (*Lablab purpureus*). The bacterium can also survive as an epiphyte on both host and non-host plants, providing a further source of initial inoculum.

Importance
Bacterial blight is a major disease of beans worldwide. It occurs only sporadically, but may be severe. Yield may be compromised if the disease develops before the pod-fill stage.

Management
- Use disease-free seed, such as certified or approved bean seed.

- Avoid moving workers and machinery between diseased and disease-free areas of the crop, particularly in wet weather.

- Use a regular protectant copper spray program to slow disease development. Copper should be tank-mixed with mancozeb for best results.

- Plough in diseased crops immediately after harvesting.

- Rotate crops with at least two years between bean crops.

- Eliminate weeds, legumes, volunteer beans and other crop hosts from production areas.

Fig 5.3 Leaf symptoms of common bacterial blight on bean.

Fig 5.4 Common bacterial blight on bean pods.

■ HALO BLIGHT

Cause
The bacterium *Pseudomonas syringae* pv. *phaseolicola*.

Symptoms
Halo blight attacks the foliage, pods and stems of bean plants.

Leaf symptoms consist of small, tan, angular, greasy spots, surrounded by a wide, diffuse, lemon-green halo. The halo is sometimes difficult to detect during hot weather. Some strains of the bacterium do not induce the chlorotic halo. The lesions expand and coalesce into larger necrotic lesions that can be 2–3 cm in diameter. Invasion of the water-conducting tissues of the plant may cause wilting, leaf yellowing, stunting and plant death. Elongated, dark green, water-soaked areas can develop on stems.

Pod symptoms consist of circular, red or brown, water-soaked lesions. A pearly white to cream, crusty bacterial ooze develops in the centre of the lesions. These 'grease spots' eventually become slightly depressed and rusty-brown in colour, and are difficult to distinguish from symptoms of bacterial blight. Infected seed, particularly those of white-seeded varieties, may show distinct yellow patches on the seed coat or, in severe cases, the whole seed may appear wrinkled.

Source of infection and spread
The bacterium is seed-borne and planting infested seed is the most important means of introducing the disease. Seed contamination arises through the vascular tissues, by direct pod infection, or by contact of the seed coat with bacteria in plant debris during harvesting and handling.

Fig 5.5 Halo blight on bean leaves.

Fig 5.6 Halo blight on bean pods.

Cool, wet weather favours infection. The disease develops most rapidly between 16 and 20°C, however warmer temperatures (20–23°C) are required for the halo symptom to develop. Once established, the pathogen spreads by wind, rain, and workers and machines moving through the wet crop. The pathogen enters plants through natural openings such as stomata or hydathodes, or through wounds.

Susceptible tropical legumes, particularly siratro (*Macroptilium atropurpureum*), glycine (*Neonotonia wightii*) and phasey bean (*Macroptilium lathyroides*) may provide a source of infection for nearby bean crops. The bacterium also may survive on undecomposed crop residues and as an epiphyte on the leaves of tolerant and susceptible bean varieties, although this mode of survival may not be as important for the halo blight bacterium as it is for other bean bacterial blights.

Importance
Halo blight is a serious disease of beans worldwide and can be particularly destructive if infested seed is planted. It occurs infrequently in Australia, but outbreaks that do occur may be severe.

Management
- Use disease-free seed, such as certified or approved bean seed.
- Avoid moving workers and machinery between diseased and disease-free areas of the crop, particularly in wet weather.
- Use a regular, protectant copper spray program to slow disease development. Copper should be tank-mixed with mancozeb for best results.
- Plough in diseased crops immediately after harvesting.

- Destroy alternative hosts of the pathogen near bean crops.

- Rotate beans with other crops.

■ POD TWIST

Cause
The bacterium *Pseudomonas flectens*.

Symptoms
Small, water-soaked areas occur on young pods, which may wither and fall. Remaining pods are curled and twisted, since infected areas fail to enlarge. Droplets of milky exudate appear on the water-soaked areas and dry to a shiny encrustation. Affected areas darken in colour.

Source of infection and spread
The main sources of the bacterium are diseased French bean, phasey bean (*Macroptilium lathyroides*) and siratro (*Macroptilium atropurpureum*). The bacterium is spread from plant to plant by the bean blossom thrip (*Taeniothrips nigricornis*). The insect alone can cause considerable twisting and scarring of pods.

Importance
Pod twist is a very minor disease in commercial bean crops that mature in the cooler months. It is more likely to occur in warmer months and in north Queensland.

Management
- Apply insecticides at flowering to control thrips.

- Plough in diseased crops immediately after harvesting.

Fig 5.7 Bacterial pod twist on bean.

FUNGI

■ ANGULAR LEAF SPOT

Cause
The fungus *Phaeoisariopsis griseola*.

Symptoms
Symptoms can occur on all aerial parts of the plant. Circular spots, up to 10 mm in diameter, develop on primary leaves and often have a zonate or target appearance. On primary leaves, lesions are generally circular rather than angular in shape. Spots on the trifoliate leaves are smaller, up to 3 mm wide, and angular. Small black clusters of bristles (synnemata) that bear the fungal spores are generally visible on the spots on the undersides of the leaves. Dark, sunken patches of varying size occur on pods and often have diffuse margins.

Fig 5.8 Angular leaf spot on a primary bean leaf.

Fig 5.9 Angular leaf spot on a trifoliate bean leaf. Note that the spots are more angular than those on primary leaves.

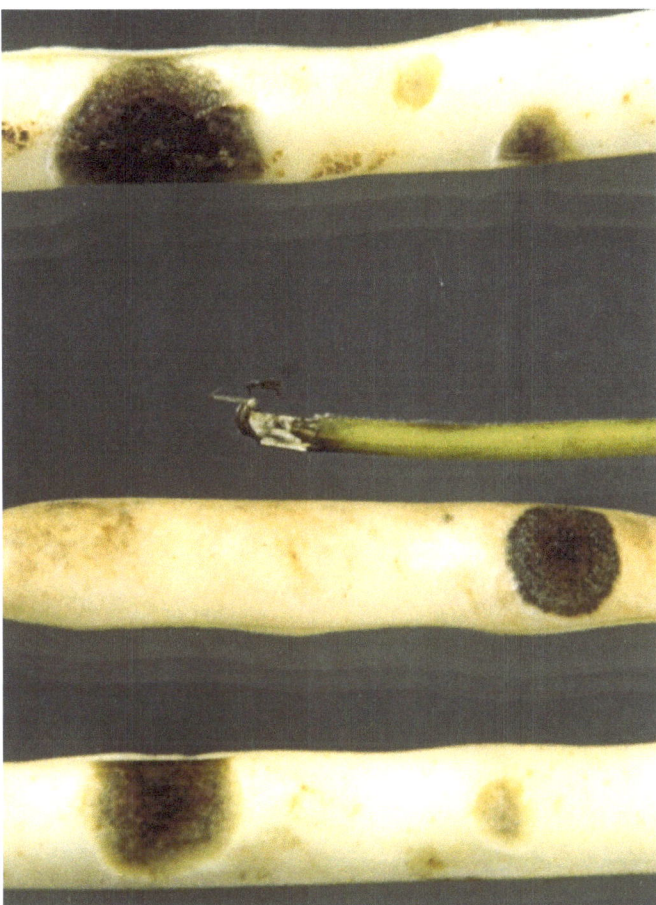

Fig 5.10 Angular leaf spot on bean pods.

Source of infection and spread

Contaminated seed is an important means of introducing the pathogen. Seed contamination results from direct pod infection or contamination of the seed coat during harvesting.

The disease spreads rapidly during wet, windy weather. The fungus also survives on undecomposed crop residues and on older, diseased crops. Infection and disease occur between 16 and 28°C, with 24°C being the optimal temperature for disease development. Humid or wet conditions promote fungus sporulation.

Importance

Angular leaf spot is usually a minor disease in well-managed crops.

Management

* Plant resistant cultivars.

* Plough in diseased crops immediately after harvest.

* Apply the recommended fungicides.

■ ANTHRACNOSE

Cause
The fungus *Colletotrichum lindemuthianum*.

Symptoms
All aerial portions of the bean plant can be affected. Angular red/brown spots develop in the leaves, with dark streaks along the veins on the undersides of leaves. Stems and petioles have dark, sunken, 1–2 cm diameter elliptical or circular lesions with red/brown margins that may girdle the stems causing the stems to collapse. Pod infection causes dark brown sunken spots, the centres of which develop pink, waxy spores during moist weather.

Source of infection and spread
Sowing contaminated seed is the most important means of introducing the pathogen. Survival on crop residues is limited, however if crop-free periods are short between crops, the fungus may carry-over on crop debris to subsequent plantings. The fungus is spread by wind-driven rain.

Importance
Anthracnose was widespread and serious before the introduction of certified and approved bean seed schemes. The disease has been practically eradicated from commercial production in Australia.

Management
* Use disease-free seed, e.g. certified or approved bean seed.

Fig 5.11 Anthracnose on bean pods.

■ ASCOCHYTA SPOT

Cause
The fungus *Phoma exigua*; syn. *Ascochyta phaseolorum*.

Symptoms
Circular, grey to brown spots develop on leaves, often marked by concentric rings with a well-defined, dark brown margin. Most spots are 6–12 mm wide, but may reach 25 mm under very favourable conditions. The centres of spots eventually dry and crack, leaving ragged holes. Many small, black fruiting bodies (pycnidia) of the fungus may be visible embedded in affected tissues.

Pod infection causes large, dark spots to occur mainly around injury sites. Infection of the floral remnants causes a dark, dry rot of the pod extending from the stem end.

Source of infection and spread
The fruiting bodies produced on diseased plants contain large numbers of spores that spread by splashing water.

Fig 5.12 Ascochyta spot on bean leaves, showing concentric rings and defined margins (left).

Fig 5.13 Ascochyta infection on bean pods.

High humidity and temperatures less than 28°C favour infection. Plants grown in locations exposed to wind are very susceptible to attack, the fungus gaining entry through damaged leaves and pods. Spots may also develop around rust pustules, insect feeding sites and other injuries.

Several tropical legumes are also hosts and the fungus may also be carried on bean seed.

Importance
Ascochyta spot is a minor disease of beans.

■ ASHY STEM BLIGHT

Cause
The fungus *Macrophomina phaseolina*.

Symptoms
On seedlings, a small, black, sunken, sharply defined lesion develops on the stem at the soil line or at the bases of the cotyledons. The infection may extend in either direction along the stem, and often extends upwards, girdling the stem and killing the growing point. In many cases, only one side of the stem is affected.

Older plants may show similar stem lesions. The lesions have a dark margin and often some concentric markings. An ashy-grey centre develops and small, black resting bodies are visible. Wilting, chlorosis and leaf death is more pronounced on one side of the plant and affected plants eventually die. The fungus may also invade roots, causing the grey-coloured tissue to breakdown. Grey lesions may also form on pods and seeds.

Source of infection and spread
The fungus survives in the soil for long periods and spreads when infested soil is moved. It may also persist in crop debris.

Fig 5.14 Ashy stem blight on bean. The affected plant on the left is wilting and the stem shows ashy grey discolouration.

The disease is favoured by hot, dry weather. Temperatures greater than 27°C in the top 2.5 cm of soil favour disease development.

Importance

Ashy stem blight is usually a minor disease in beans in Australia.

Management

- Only plant high quality, disease-free seed.

- Deep-plough old crops and ensure the residues have completely broken down before planting a subsequent crop.

- Plant bean crops on soil with good moisture and maintain optimum growing conditions.

■ CERCOSPORA LEAF SPOT

Cause

The fungus *Cercospora canescens*.

Symptoms

Infected leaves, particularly those that are more mature, develop circular or slightly angular, greyish spots, sometimes with reddish margins. Spots may dry and portions may fall out, giving the leaf a ragged appearance.

Source of infection and spread

The fungus may be a seed contaminant or survive in crop debris. Spores spread by wind.

Importance

Cercospora leaf spot is a minor disease, usually seen only on older, senescing leaves.

Fig 5.15 Cercospora leaf spot on the upper and lower surfaces of bean leaves.

■ COTTONY LEAK AND STEM ROT

Cause

The oomycete *Pythium aphanidermatum*.

Symptoms

Soft, water-soaked areas develop on leaves and stems, often covered in a fine, white, cottony growth in wet weather. The pathogen may also cause seedlings to damp off and root rot in mature plants.

Watery, soft rot can develop on pods in transit or storage, or abundant, white, cottony growth mats pods together into 'nests' that later become a soft, leaking mass known as 'cottony leak'.

Source of infection and spread

The pathogen is a common soil inhabitant, surviving in organic matter and as resistant oospores. Soil moisture and temperature have a major influence in disease development and severity with warm, wet weather favouring disease development.

Importance

Generally, cottony leak is a minor disease, but serious losses can occur in hot, wet weather. The stem rot phase may cause thinning of stands.

Management

- For stem rot in the field:

  - Avoid close planting and sowing seed deep in the soil seed.

  - Cultivate carefully to avoid plant injury.

  - Do not plant in poorly drained areas.

  - Prepare land thoroughly to allow residues to decompose.

Fig 5.16 Cottony leak caused by *Pythium* on bean pods.

- For cottony leak:
  - Discard diseased pods and pack only dry pods.
  - Store in well-ventilated areas at temperatures between 12 and 15°C.

■ PLEIOCHAETA BROWN SPOT

Cause
The fungus *Pleiochaeta setosa*.

Symptoms
Leaf symptoms consist of reddish-brown spots rarely more than 2 mm wide. With age, the centres eventually fall out, leaving ragged holes. Small, dark spots often develop on veins on the underside of the leaves.

Slightly sunken spots with dark centres and light brown margins occur on pods. Spots are up to 2 mm wide and may coalesce.

Fig 5.18 Pleiochaeta brown spot symptoms on bean pods.

Source of infection and spread
This disease is restricted to areas where beans grow on very light, sandy soils. The fungus is a wound pathogen only, and leaf and pod abrasion by sand during strong winds predisposes plants to attack.

Streaked rattlepod (*Crotalaria pallida*) and Gambia pea (*Crotalaria goreensis*) are alternative hosts of the fungus.

Importance
The disease is an occasional problem in wind-damaged crops on sandy soils.

Management
- Eradicate alternative weed hosts.
- Establish windbreaks in exposed locations.

■ POWDERY MILDEW

Cause
The fungus *Erysiphe polygoni*.

Symptoms
Darkened, mottled spots develop on the upper surface of the older leaves. The spots become covered by a circular

Fig 5.17 Pleiochaeta brown spot on bean leaves.

Fig 5.19 Powdery mildew symptoms on bean stems and the underside of leaves.

Fig 5.20 *Rhizoctonia* infection of bean roots (below).

growth of white, superficial powdery mycelium. Severe infection may cause the leaves to become distorted and yellow. Stems and pods may also be affected.

Source of infection and spread

The fungus survives on crop residues and volunteer plants and disperses as air-borne spores. Disease development is favoured by warm, dry days and cool nights with morning dews.

Importance

Powdery mildew usually occurs late in the season and causes little damage. Infection during flowering and pod set may require specific control measures.

Management

• Apply recommended fungicides if disease severity and crop growth stage warrant action.

■ RHIZOCTONIA ROT

Cause

The fungus *Rhizoctonia solani*.

Symptoms

Sunken, brick-red lesions occur on the lower stem and roots, often before seedlings emerge. Some plants are killed and others remain stunted. Many plants recover to give a satisfactory crop after producing new roots above the diseased area.

Reddish spots occur where pods contact infected soil. The fungus spreads rapidly during transit, often with a brown fungal growth appearing on affected areas.

Source of infection and spread

The fungus is a common soil inhabitant affecting a wide range of plants. The disease is often severe in areas where large amounts of plant residue remain in the soil. Bean hypocotyls are most at risk of infection in the first two weeks after planting.

Importance

Usually, rhizoctonia rot is more important in crops grown on light soils where seedling loss and pod infection can be serious.

Management

• Prepare land thoroughly so that plant residues are completely broken down before planting.

• Apply the registered soil fungicide to blocks prior to planting.

• Hill plants to encourage new root growth.

Fig 5.21 *Rhizoctonia* on bean seedlings.

■ ROOT ROT COMPLEX

Cause

The pathogens *Fusarium solani*, *Aphanomyces euteiches*, *Pythium* spp., *Thielaviopsis basicola* and *Rhizoctonia solani*.

Symptoms

Plants are yellow and stunted; the crowns have a red to brown-black discolouration and the taproot dries and may be destroyed. Plants often produce a cluster of fibrous roots just below ground level and, if growing conditions are good, may recover and yield satisfactorily. Affected plants are commonly unthrifty or stunted and in advanced cases will wither and die.

Source of infection and spread

The pathogens causing root rot are soil inhabitants and disease may develop over a wide range of temperatures, depending on which organisms are present. Disease frequently develops in plants grown under stress. Low soil temperatures at planting, deep sowing and wet conditions during growth favour infection by *Fusarium solani*, *Aphanomyces euteiches* and *Pythium* spp. Wet, poorly drained soils and high temperatures are conducive to infection by *Theilaviopsis basicola* and some *Pythium* species. Disease is more severe in compacted soils and where legumes have been grown continuously for many years.

Importance

Root rot can cause serious losses, particularly when extremes of temperature and high soil moisture conditions prevail and where beans are cropped continuously.

Fig 5.23 Stem symptoms of *Thielaviopsis basicola* infection (bean root rot complex).

Management

- Deep-rip soils to improve drainage and aid root penetration.

- Avoid deep planting; sow at less than 25 mm during the winter months.

Fig 5.22 Bean root rot complex is caused by several soil-borne pathogens.

Fig 5.24 A comparison of healthy and rotted root systems (bean root rot complex).

- On light soils, hilling will encourage new root growth, although this is not recommended on heavy soils.

- Deep-plough old crops and ensure residues have completely broken down before planting a subsequent crop.

- Rotate beans with crops other than legumes.

■ RUST

Cause
The fungus *Uromyces appendiculatus*.

Symptoms
The common symptom of bean rust is scattered, reddish-brown, circular pustules on leaves or pods. Pustules, at first light green in colour, rupture the epidermis to produce abundant, powdery, reddish-brown spores (urediniospores). The pustules vary in size from pinpoint to 1 or 2 mm in diameter. Large pustules often surrounded by a halo of yellow host tissue. During cold weather, spore masses may be black rather than red. Pustules may also develop on pods, but stems are not usually affected.

Source of infection and spread
Abundant spores produced on leaves spread in wind and air movement, and initiate new infections within seven to 14 days. The dark-coloured spores are thick-walled and survive for longer periods. Infection is favoured by cool, damp weather; fogs, mists and dews provide ideal conditions.

The disease is most severe in areas where old, diseased crops remain after harvesting, providing inoculum to infect younger crops. The bean rust fungus is highly variable genetically, and there are several races that differ in their ability to attack bean varieties.

Fig 5.26 Rust infection on a bean leaf and pod.

Importance
Rust is only a sporadic problem but significant outbreaks may develop in cool, damp or foggy weather.

Management
- Spray with the recommended fungicides. Base the spray program on regular protectant fungicide applications, using targeted, systemic fungicide applications when conditions favour infection. Fungicide resistance studies indicate that the triazole group (Group C) of fungicides control Queensland pathogen populations less readily.

- Plough in crops immediately after harvest.

- Use resistant varieties.

Fig 5.25 Rust infection on the underside (left) and upper surface of a bean leaf.

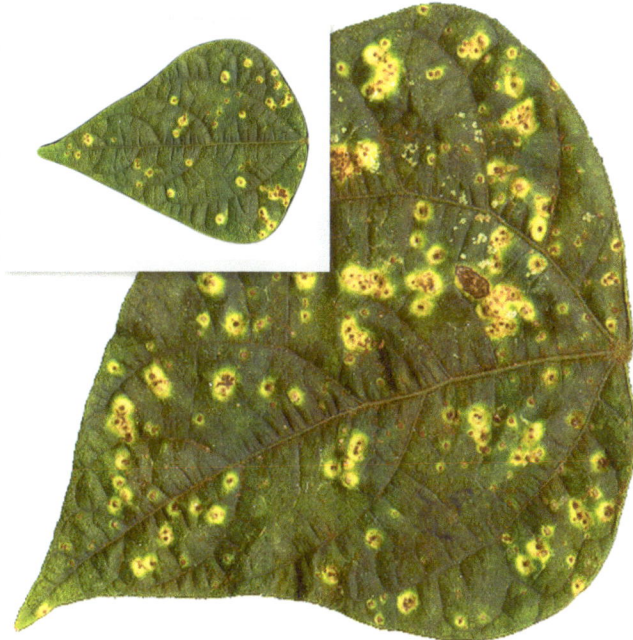

Fig 5.27 Rust symptoms on the upper side of a bean leaf.

■ SCLEROTINIA ROT

Cause

The fungus *Sclerotinia sclerotiorum*.

Refer to the chapter on Common diseases of vegetable crops.

Fig 5.30 Close view of sclerotinia 'nests' in bean pods.

Fig 5.28 Sclerotinia rot causing wilting and plant death.

NEMATODES

■ ROOT-KNOT NEMATODE

Refer to the chapter on Common diseases of vegetable crops.

VIRUSES

■ BEAN SUMMER DEATH

Cause

Tobacco yellow dwarf virus (Mastrevirus).

Symptoms

In susceptible varieties young seedlings may suddenly collapse and die following one to two weeks of hot weather. Plant death is rapid during hot weather, but in cooler weather, plants are stunted and wilted, and the younger leaves curl downward. Vascular discolouration of the roots and lower stem occurs; in advanced stages, the roots may blacken and rot.

Source of infection and spread

Natural hosts of the virus include tobacco, thornapple (*Datura* sp.), tomato and bean. The virus is transmitted by the leafhopper *Orosius argentatus* in a persistent manner.

Epidemics usually occur in inland areas of Australia following hot weather, which favours migration of the leafhopper vector into crops of susceptible bean varieties.

Fig 5.29 Sclerotinia 'nests' in bean pods.

Fig 5.31 Bean summer death showing the susceptible variety on the left.

Importance

Bean varieties differ considerably in their reaction to the virus. Many stringless varieties are highly susceptible while many stringed and dry bean varieties are resistant.

Sporadic epidemics can cause complete loss of crops of susceptible varieties.

Management

• Plant resistant varieties when conditions are likely to favour virus infection.

■ BEAN YELLOW MOSAIC

Cause

Bean yellow mosaic virus (Potyvirus).

Symptoms

Leaves develop a mosaic pattern of contrasting dark and yellowish-green areas, often accompanied by bright yellow spots. Severe strains may cause rough, wrinkled, malformed leaves.

Source of infection and spread

The virus is spread by aphids, and only short feeding periods are required for transmission to take place. Other hosts include clovers, medics and gladioli.

Importance

Usually, bean yellow mosaic is a minor disease in Australia and control measures are generally not warranted.

■ COMMON MOSAIC

Cause

Bean common mosaic virus (Potyvirus).

Symptoms

The virus causes cupping and twisting in leaves, which show a light green and dark green mosaic pattern. The dark green tissue is often bubbled and/or in bands next to the veins. Affected plants produce smaller, curled pods with a greasy appearance, and reduced yields.

Source of infection and spread

The virus spreads between production areas primarily in infected bean seed, which also maintains the virus between seasons. During the growing season, aphids spread the virus. Navy beans and some other legumes are also hosts of the virus.

Fig 5.32 *Bean yellow mosaic virus.*

Fig 5.33 Symptoms of *Bean common mosaic virus.*

Importance

Common mosaic can be serious when virus-infected seed is planted.

Management

- Use disease-free seed, e.g. certified or approved bean seed.

- Plant resistant varieties.

■ PEANUT MOTTLE

Cause

Peanut mottle virus (Potyvirus).

Symptoms

Peanut mottle virus infects both French and navy bean varieties. The symptoms produced vary greatly, depending on the variety involved. Some varieties, e.g. Labrador, Canyon and Gresham, are resistant and do not develop symptoms in the field. Varieties such as Kerman and Gallaroy navy beans are moderately susceptible, developing leaf mottling and some leaf distortion. The most severe response to *Peanut mottle virus* infection is a severe, systemic necrotic reaction resulting in death or severe stunting. The terminal leaflets often wither and die, with extensive necrosis occurring on the stem, petiole and leaf veins. Leaves often appear twisted and frequently fall from the plant. A spreading network of necrotic leaf veins develops from the point of infection.

Pods may show dark, oily spots or a reddish-brown necrosis. Immature pods frequently shrivel and dry out.

Source of infection and spread

The natural hosts of *Peanut mottle virus* in Australia are peanut, French and navy bean, soybean, pea, mung bean

Fig 5.34 *Peanut mottle virus* in a highly susceptible bean variety.

Fig 5.35 *Peanut mottle virus* on leaves.

and some *Cassia* species. The virus spreads from plant to plant by aphids, which require only very brief feeding periods for transmission. *Peanut mottle virus* is also carried in the seed of several hosts, particularly peanut. Virus-infected peanut seed usually provides the initial source of the virus, and aphids then spread the virus within peanuts and into other hosts. There is a low percentage of seed transmission in navy beans, but most infection in beans results from aphids moving from adjacent peanut crops containing diseased plants.

Importance

Peanut mottle can be serious in crops of susceptible navy bean varieties growing near peanuts. The virus is only of minor importance in French beans.

Management

- Avoid planting bean varieties susceptible to peanut mottle in areas where peanuts are grown.

- Use the available resistant varieties.

■ SUB CLOVER STUNT

Cause

Subterranean clover stunt virus (Nanovirus)

Symptoms

Leaves curl downward, yellow and thicken. The plant becomes stunted and the growing point often dies. The few pods that may set are small and distorted.

Source of infection and spread

The virus is spread by aphids, particularly the cowpea aphid (*Aphis craccivora*). The natural hosts of the virus are confined to legumes and include subterranean clover (*Trifolium subterranean*), other clover species, medics (*Medicago* species), broad bean, pea and bean.

Aphids migrating from susceptible pasture species often transmit the virus to susceptible bean varieties.

Importance

The virus can cause serious losses in susceptible bean varieties in parts of southern Australia, particularly during spring when migration of the aphid vector often occurs.

Management

- Plant resistant varieties when the disease is likely to be a problem.

■ FURTHER INFORMATION

Duff J (2008). *Green beans: insect pests, beneficials and diseases.* Department of Primary Industries and Fisheries, Brisbane, Queensland.

Schwartz HF, Steadman JR, Hall R & Forster RL (Eds) (2005). *Compendium of bean diseases* (2nd edn). APS Press, St Paul, Minnesota.

Beetroot (*Beta vulgaris*) and silver beet (*Beta vulgaris* var. *sicla*) are members of family Amaranthaceae and their centre of origin is the Mediterranean and North Africa. Beetroot is grown for its swollen, edible root and young leaves, which are used as a leafy salad vegetable. Silver beet produces large, edible leaves with prominent leaf stalks. Both crops are propagated by direct seeding.

Beetroot is a cool temperate crop, grown during the cooler months in south Queensland and throughout the year in cooler regions.

A large proportion of the beetroot crop is grown in southern Queensland and processed as canned beetroot.

Poor crop establishment related to soil borne pathogens is a major industry problem.

BACTERIA

■ BACTERIAL BLIGHT

Cause
The bacterium *Pseudomonas syringae* pv. *aptata*.

Symptoms
Initially, dark spots form on the leaf blades, and later become large irregular lesions with pale centres. Lesions also develop around leaf margins and on midribs. A dark soft rot may develop on the petioles.

Source of infection and spread
The primary source of infection is contaminated seed. The pathogen is spread in irrigation water and by wind-driven rain and water splash.

Importance
Bacterial blight is an uncommon disease in Australia.

Management
• Foliar copper sprays help to limit secondary spread of the disease

FUNGI

■ ROOT ROT

Cause
A complex of soil-borne pathogens including *Pythium* spp. (*P. aphanidermatum*, *P. ultimum*, and *P. dissotocum*), *Rhizoctonia solani* and *Aphanomyces cochlioides*.

Symptoms
In beetroot, root rot may occur in plants soon after emergence, or in more mature plants, during root expansion. In young plants, a dark brown to black discolouration develops on the developing taproot. Severely affected plants may wither and die at the seedling stage.

Fig 6.1 Leaf symptoms of bacterial blight.

Fig 6.2 Symptoms of black root rot on seedlings.

Fig 6.3 Collapsed young beetroot plants affected by root rot.

Plants that survive early disease infection may develop misshapen roots. Leaves on affected plants commonly turn red.

In older plants, large, roughly circular, depressed, dark lesions occur on the globes. Excessive proliferation of secondary feeder roots is a sign of *Pythium* infection. Above-ground symptoms include general plant wilting and yellowing and death of the oldest leaves.

Source of infection and spread

The pathogens are soil-borne and thrive under wet conditions. Infection and disease development are favoured by warm weather and high soil moisture.

Very young plants are the most susceptible to infection by *Pythium* spp. *P. aphanidermatum* is the most pathogenic of the three main species of *Pythium* affecting beetroot. It

Fig 6.4 Wilting in silver beet plants caused by root rot infection.

Fig 6.5 Field symptoms of plants in a waterlogged soil affected by *Aphanomyces cochlioides*.

causes disease at temperatures greater than 15°C in newly emerged seedlings, but higher temperatures (30–35°C) are required before infection will cause losses in older plants. *P. dissotocum* causes disease at temperures greater than 30°C. *Pythium* spreads through the soil as swimming zoospores. Thick-walled oospores enable its long-term survival in infected soil.

Aphanomyces cochlioides is similar in its environmental requirements to *Pythium*. In wet soils and in high temperatures (20–28°C), plant infections are initiated by zoospores, and oospores enable long-term survival of the disease, even when the soil dries out.

Rhizoctonia solani survives as hyphae in colonised host tissues. Under warm soil temperatures (25–33°C) the fungus grows through the soil and infects the plant through its leaves, petioles, crown and roots. *Rhizoctonia* root rot occurs in most types of soil, but is most severe in heavy, poorly drained soils, especially in low areas where water collects.

Fig 6.6 *Aphanomyces cochlioides* causes depressed, dark lesions on beetroot globes.

Importance
Root rots often causes serious losses in poorly drained areas and after periods of prolonged, wet weather following planting. Root rots have become a substantial problem for the processing beetroot industry in early and very late plantings, when crops are grown under conditions highly favourable to the root rot pathogens.

Management
- Avoid poorly drained areas.
- Provide good seedbed tilth.
- Plant into raised seedbeds.
- Apply the recommended fungicide seed treatments
- Dolichos or barley rotations should be used between beet crops to prevent *Pythium* and *Rhizoctonia* build-up.
- Ensure all crop residues are completely decomposed before re-plant to help limit re-infection by *Rhizoctonia*.

- Plant beet crops when the risk of hot, wet soils is minimal. In Queensland, crops planted before April are most at risk of infection.

■ DOWNY MILDEW

Cause
The oomycete *Peronospora farinosa*.

Symptoms
Initial symptoms are chlorosis and distortion of the youngest leaves. The pathogen grows systemically through the leaves and invades the growing point causing the leaves to become thickened and malformed. Chlorotic, irregularly shaped lesions develop on the leaves. In high humidity, grey, mycelial growth may appear on the underside of the lesions. Lesions later dry out and turn brown.

Source of infection and spread
The pathogen can be seed-borne. Thick-walled oospores can survive in the soil and germinate to produce mycelium and sporangia. Air-borne sporangia are carried by wind to susceptible host plants. Cool, moist conditions are required for epidemics to develop. At least six hours of leaf wetness and cool temperatures (7–15°C) are required for infection to occur. Temperatures above 20°C inhibit infection.

Importance
Downy mildew is a minor disease of beet crops in Australia.

Management
- Apply the recommended chemicals, particularly during cool, wet weather.

■ LEAF SPOT

Cause
The fungi *Cercospora beticola* and *Phoma betae*.

Symptoms
Cercospora leaf spot: Small, brown flecks develop with a reddish border, expanding to circular spots about 4 mm wide with an ashy-grey centre. This tissue becomes thin and brittle, and often drops out, leaving a ragged hole.

Phoma leaf spot: Large, light brown spots, often with concentric markings, occur on leaves. Numerous, glistening black fruiting bodies (pycnidia) form on mature spots.

Fig 6.7 Cercospora leaf spot on beetroot.

Fig 6.8 Cercospora lesions on silver beet leaf and stem.

Fig 6.9 Close-up view of a Phoma leaf spot lesion on a beetroot leaf.

Source of infection and spread

The fungi survive on undecomposed beet residues in the soil, on weed hosts and on beet seed. Hosts include beetroot, silver beet, sugarbeet, spinach and several *Atriplex* and *Chenopodium* weed species. Leaf spot is favoured by warm, wet weather. Severe outbreaks generally require a period of showery weather.

Importance

Cercospora leaf spot is a common disease in beetroot and silver beet but is usually unimportant in well-managed crops. It may be a significant problem in crops grown for baby-leaf production, because the foliage is the saleable product.

Management

- Apply the recommended fungicides, particularly during warm, wet weather.

- Rotate beet crops with other non-host vegetables.

- Control weeds, particularly *Chenopodium* weeds like fat-hen, in and around beet crops.

Fig 6.10 Phoma leaf spot on silverbeet.

■ POWDERY MILDEW

Cause
The fungus *Erysiphe polygoni*.

Symptoms
Whitish-grey, superficial spots develop first on the older leaves and occur on both upper and lower leaf surfaces. The spots enlarge and leaves become covered by greyish fungal growth. Older leaves that are severely affected may become chlorotic and then senesce.

Source of infection and spread
Spores of the fungus are spread by wind and may be transported at high altitude over long distances. Germination of the spores occurs over a wide range of environmental conditions, optimally at 25°C and 70–100% relative humidity. Epidemics occur during dry weather, alternating with periods of high relative humidity and temperatures greater than 20°C. Plant susceptibility increases with age, and damage is most severe on drought-stressed plants. Dry, warm weather favours the development of the disease. Powdery mildew carries over between crops in buried roots and on volunteer beet and weed hosts.

Importance
Powdery mildew is often a severe disease in silver beet.

Management
- Destroy old crops promptly.
- Apply the recommended fungicides.

NEMATODES

■ BEET CYST NEMATODE

Cause
Beet cyst nematode (*Heterodera schachtii*).

Fig 6.11 Powdery mildew on silver beet.

Fig 6.12 Leaf spotting caused by powdery mildew on a beetroot leaf.

Fig 6.13 Beet cyst nematode. Note the development of fine feeder roots and small white cysts on the roots.

Symptoms

Affected plants appear stunted, often developing large numbers of fine 'feeder' roots. Female nematodes are visible as small, white cysts on the plant roots. The next generation develops as eggs within the female body, which changes to a brown colour as it develops into a mature cyst. The presence of cysts on the roots distinguishes cyst nematode from root-knot nematode that causes galls on roots.

Source of infection and spread

The nematode survives in soil for many years as the egg stage within the cyst. The cysts disperse with soil, in run-off water, and on farm machinery, workers and animals. The nematode can also reproduce on brassica species, rhubarb, spinach and turnip.

Importance

The beet cyst nematode has been found in several Australian states on beetroot, brassicas and rhubarb.

Management

- Do not plant beetroot or brassicas on infested land for at least five years.

- Treat soil before planting with a recommended nematicide.

- Clean farm machinery and packing crates before moving them from an infested area.

■ ROOT-KNOT NEMATODE

Refer to the chapter on Common diseases of vegetable crops.

Fig 6.14 Root-knot nematode on beetroot.

■ FURTHER INFORMATION

Harveson RM, Hanson LE, Hein GL (Eds) (2009). *Compendium of beet diseases and pests* (2nd edn). APS Press, St Paul, Minnesota.

Members of the Brassicaceae or mustard family are grown worldwide as vegetable, oilseed or fodder crops. The major brassica vegetable crops are classified as *Brassica oleracea* and include cabbage, cauliflower, broccoli and Brussels sprouts. These are often called cole crops. *Brassica rapa* includes the leafy and root vegetables Chinese cabbage, pak choy and turnip. The oilseed crop canola, widely grown in southern Australia, is *Brassica napus*.

Radish (*Rhaphanus sativus*) is a root crop while other root crops within the family include horseradish, watercress, cress and rocket salad.

Brassica crops are grown either by direct seeding or by transplanting seedling plants produced in nurseries into production fields.

BACTERIA

■ BLACK ROT AND LEAF SCALD

Cause
The bacterium *Xanthomonas campestris* pv. *campestris*.

Symptoms
The first symptoms to develop are usually small, water-soaked spots on the leaf margins. These expand towards the midrib exhibiting characteristic, yellow, V-shaped areas with a broad yellow margin and a grey brown centre. Vein blackening may extend down the leaves into the petiole and the stem. A cut across the stem reveals a black ring in the water-conducting tissues just beneath the bark.

Development of the V-shaped marginal lesions occurs when the bacteria enter the leaf in water through hydathodes, natural openings at the vein ends on leaf margins. When bacteria enter the vascular system directly through wounds, root injury or hail damage, the leaf blade may yellow and the leaf collapse before the symptoms on leaf margins develop.

Soft-rotting organisms often quickly follow the invasion of fleshy leaves and petioles by the bacterium causing black rot, causing the plants to rot rapidly.

Certain strains of the bacterium infect plants through the leaf stomata (breathing pores) causing scald symptoms. Tan, circular spots with yellow haloes develop between the

Fig 7.1 Black rot of cauliflower (bottom) and broccoli.

Fig 7.2 Black rot of cabbage. Note the V-shaped lesions extending from the leaf margins.

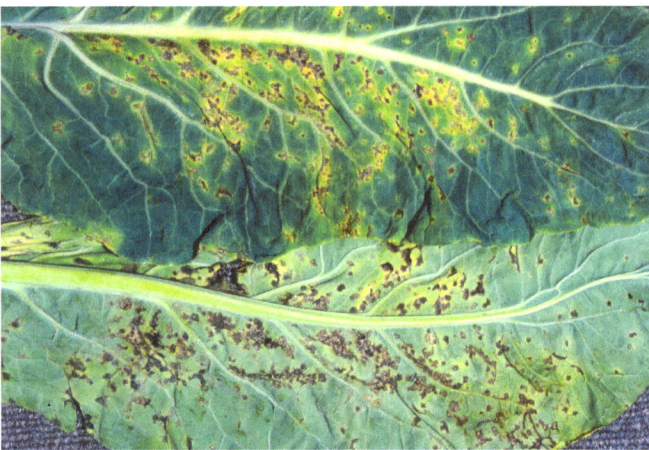

Fig 7.3 Leaf symptoms of black rot on upper (top) and lower leaf surfaces. Note the small, grey-brown, water-soaked lesions and leaf yellowing.

veins. These coalesce and dry out to give the leaf scald symptoms. Affected areas often crack and disintegrate, giving the leaf a tattered appearance.

Source of infection and spread

The bacterium may be introduced in seed with the pathogen also surviving in undecomposed crop residues, brassica weeds and other brassica crops. Bacteria spread in water splash during wet, windy weather or by overhead irrigation. The pathogen also disperses on insects and on people and equipment moving through a crop. Warm, humid weather favours rapid disease development.

Importance

Black rot is one of the most serious diseases of brassicas worldwide, particularly in warm, humid climates.

Management

- Use seed treated in hot water to reduce seed-borne infection. Treat cabbage and Brussels sprouts at 50°C for 25 minutes; cauliflower and broccoli at 50°C for 25 minutes. Keep seedling production areas separate from field production areas.

- Prevent overcrowding of seedlings.

- Keep seedling production areas free from susceptible weeds.

- Plough in diseased crops immediately after harvesting. In each field rotate crops so that brassicas are grown only once in every three crops.

- Control weeds and insects.

- Apply a regular, preventative spray program of copper to limit the spread of black rot, particularly in seedlings. For best results, copper should be tank-mixed with mancozeb.

■ HEAD ROT

Cause

Several species of bacteria, including *Erwinia* spp. and *Pseudomonas* spp.

Symptoms

A slimy soft rot develops on heads of broccoli and cauliflower. An unpleasant odour is often associated with affected heads.

Source of infection and spread

The bacteria occur with decaying organic matter in the soil. Bacteria often invade heads through injury sites. Hot weather favours the development of rot, which may be very severe if wet weather occurs. In the field, spread is by water splash and on cutting knives. The bacteria can spread by contact during transit and storage. Severe head rot appears to be associated with high application rates of nitrogen fertiliser.

Fig 7.4 Upper and lower leaf symptoms of black rot.

Fig 7.5 Bacterial head rot of broccoli.

Fig 7.6 Peppery leaf spot of Brussels sprouts.

Importance

Head rot is often a major cause of loss in broccoli and cauliflower, particularly if wet weather occurs during the late summer production period.

Management

- Avoid harvesting when crops are wet.
- Avoid high rates of nitrogen fertiliser application.
- Sterilise cutting knives if the disease occurs in the field.
- Remove field heat from heads as quickly as possible after harvest and store at 0°C and 95% relative humidity.
- Grow less susceptible varieties.

■ PEPPERY LEAF SPOT (BACTERIAL LEAF SPOT)

Cause

The bacterium *Pseudomonas syringae* pv. *maculicola*.

Symptoms

A general flecking appearance of leaves results from small (1 mm diameter), purple to brown spots surrounded by chlorotic haloes. Under wet conditions, spots may coalesce to cover large areas of the leaves. Infection of the veins may cause leaf puckering.

Source of infection and spread

The bacterium is seed-borne and survives in crop residues. It disperses in water splash and wind-driven rain. The disease is favoured by cool (22–23°C is optimal), wet weather.

Importance

Cauliflower is the main host but cabbage, broccoli, Brussels sprouts and other brassicas are also affected. Occasional, severe losses can occur following frost or cold, wet weather.

Management

Refer to the descriptions of black rot and leaf scald.

■ SOFT ROT

Cause

The bacterium *Erwinia carotovora*.

Refer to the chapter on Common diseases of vegetable crops.

■ ZONATE LEAF SPOT

Cause

The bacterium *Pseudomonas cichorii*.

Symptoms

Circular to irregular, water-soaked, zonate spots develop on leaves. The spots are at first light brown but turn darker as they mature. If the leaves of the head are pulled away, irregular spotting and necrotic areas are usually visible well into the head.

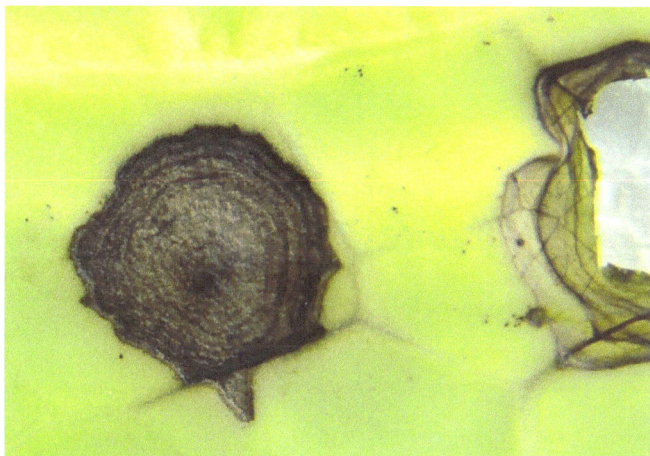

Fig 7.7 Zonate leaf spot of cabbage.

Fig 7.8 Detailed view of zonate leaf spot of cabbage.

Source of infection and spread

The bacterium survives on undecomposed crop residues and may also commonly inhabit soils. Rapid spread occurs during wet, windy weather or by overhead irrigation. The bacterium has a wide range of hosts including lettuce, endive, clover, chicory and Iceland poppy.

Importance

Zonate leaf spot is a minor disease on cabbage.

Management

• Avoid highly susceptible varieties.

FUNGI

■ ALTERNARIA SPOT

Cause

The fungus *Alternaria brassicicola*.

Symptoms

Small, dark specks occur on leaves and later enlarge into circular dark grey-brown to black spreading spots or lesions. These spots can grow to 25 mm in diameter and usually contain concentric rings, giving the spots a target-like appearance. The spots usually have a distinct margin and a sunken centre surrounded by a yellow, chlorotic halo. The spots become thin, dry and papery with age and may tear to give a 'shot hole' appearance.

Lesions may also develop on stems and petioles, flower stalks and seed heads. Dark, sunken areas may also develop in the curd of cauliflowers and on broccoli heads. The fungus may also cause seedling blight.

Fig 7.10 Leaf symptoms of Alternaria.

Source of infection and spread

The fungus is seed-borne and also survives in old plantings and crop debris. Spores spread in wind or water splash. The fungus requires six to eight hours of leaf wetness for infection and warm (20–30°C) and moist weather favours disease development.

Importance

Alternaria spot is a common disease of brassica crops and can cause a considerable reduction in quality and appearance when conditions are favourable.

Management

• A regular protectant fungicide program should be used, particularly when conditions are favourable for infection.

• Hot-water treatment of seed (50°C for 30 minutes) is effective against the fungus.

• Rotate crops and destroy crop residues as soon as possible after harvest.

Fig 7.9 Alternaria spot on cabbage.

■ BLACK LEG

Cause
The fungus *Leptosphaeria maculans* (*Phoma lingam*).

Symptoms
Plants can be infected at any stage of growth. Symptoms may occur on cotyledons or the first true leaves of seedlings, resulting in damping-off if the infection is severe. On leaves, grey to brown leaf spots develop and often contain numerous, tiny fungal fruiting bodies or pycnidia. The pycnidia can exude a pink spore mass under warm, humid conditions.

On stems, large, brown to black sunken cankers occur near the base of the stem, eventually girdling it and causing stunting and wilting of plants. Pycnidia of the fungus may appear on the affected area. Internally, the stem shows a brown, dry rot and the stem's woody tissues become black.

Source of infection and spread
The fungus is commonly introduced in infected seed and survives on undecomposed crop residues and in soil for several years. Low levels of seed infection (<1%) are sufficient to cause serious disease outbreaks. Large numbers of spores produced in pycnidia spread in rain or irrigation. The fungus may also spread by air borne ascospores produced on infested crop residues. Wet, windy weather with temperatures of 15–20°C favour disease development.

Importance
Black leg is an important disease of vegetable and oil-seed brassica crops worldwide. It is more common and damaging in cooler climates.

Management
• Use seed that treated in hot water (50°C for 25 minutes).

Fig 7.11 Field symptoms of black leg.

Fig 7.12 Stem symptoms of black leg.

• Produce seedling transplants away from production fields and maintain high standards of nursery practice and hygiene

• Plough in diseased crops immediately after harvest and ensure that all residues have thoroughly decomposed before replanting with brassicas.

• Rotate crops so that brassicas are not grown on the same land more than once every three to four years.

■ BLACK ROOT OF RADISH

Cause
The oomycete *Aphanomyces raphani*.

Symptoms
Dark, irregular patches develop on the main root and often coalesce to cover the root completely. Longitudinal cracking of the affected areas may follow, but the tissues do not break down unless invaded by secondary organisms.

Fig 7.13 Black rot on radish.

Source of infection and spread

The pathogen can persist in the soils for long periods as resistant resting spores. Zoospores are produced which invade the main root where the secondary roots emerge. Warm weather and moist soil favour the disease.

Importance

Black root affects radish in most production areas worldwide. It is more severe on long-rooted, white-fleshed types and less severe on coloured or late maturing varieties. The pathogen can also occur on other brassica species. In seedlings, roots, hypocotyls, petioles and cotyledons blacken and collapse.

Management

- Improve soil drainage in affected areas.

- Rotate radish with other non-brassica species.

- Grow resistant varieties.

■ CLUB ROOT

Cause

Plasmodiophora brassicae.

Fig 7.14 Field symptoms of club root.

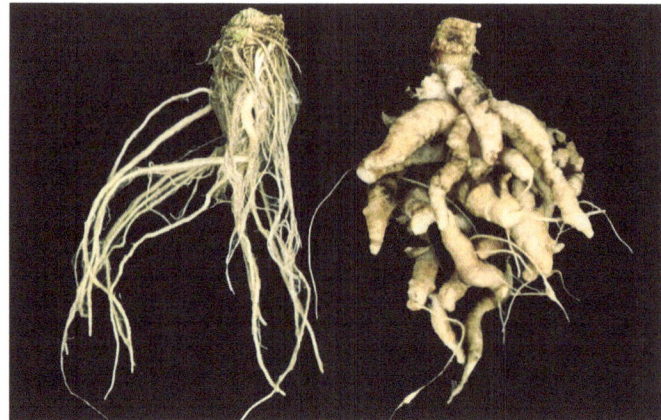

Fig 7.15 Club root symptoms (right) compared with a healthy root.

Symptoms

Infected plants are stunted, have pale green to yellowish leaves, and often wilt during the hotter part of the day. The most characteristic symptom occurs on the roots, where small or large spindle-shaped, knobby or club-shaped swellings occur. The entire root system may be affected, with older, severely affected roots disintegrating before the end of the season.

Source of infection and spread

The pathogen is an obligate parasite and can only reproduce in the root tissue of the host. It can survive in soil for decades as thick-walled resting spores that resist microbial degradation. Most resting spores occur close to the soil surface. The resting spores germinate in the presence of plant roots under warm conditions (20–25°C), producing motile zoospores, which move in soil moisture and invade root hairs and encyst. In the second stage of the life cycle, a plasmodium forms producing secondary zoospores. These may infect other root hairs or fuse to form larger cells that penetrate deeper into the internal root tissues. Infection causes root cells to multiply and enlarge, leading to the formation of galls about three to four weeks after infection. When the galls decay, large numbers of resting spores are released into the soil.

The development of club root is favoured by high soil moisture, warm temperatures (20–25°C), and significant numbers of spores in the soil to initiate infections. Although the pathogen is present in a range of soil types, disease development is favoured by acid soils (below pH 7.0).

The pathogen infects a wide range of brassica species including cabbage, cauliflower, broccoli, Brussels sprouts, turnip, radish, mustard and canola. It also infects a range of brassica weed species.

Spores of the organism disperse in soil and water. Examples include contaminated soil on machinery, bins, transplant trays, wind-blown soil, water run-off from infected paddocks, contaminated dam water used for irrigation and water movement in fields.

Importance

Club root occurs in brassica crops worldwide and can cause important production losses in warm temperate climates.

Management

- Implement good nursery and farm hygiene.

- Purchase high quality seedlings from a reputable source.

- Ensure equipment entering the property is thoroughly cleaned with a quaternary ammonium type sanitiser.

- Ensure water used for irrigation and washing is not contaminated with the club root organism.

- Cultural controls:

  - Avoid overwatering and improve drainage by increasing bed heights.

  - Rotate brassicas with other non-brassica crops using a three-year rotation.

  - Avoid double cropping and summer cropping brassicas in infested soils.

  - Control brassica weeds to reduce pathogen build up.

  - Use tolerant or resistant varieties.

- Adjust soil pH to maintain it at pH 7.0–7.5.

- Increase soil concentrations of calcium and boron to inhibit clubroot.

- Apply recommended chemicals in heavily infested sites as indicated by a disease risk assessment. Chemical efficacy is enhanced by targeting the plant root zone, where protection is needed, and applying to the transplant row immediately before planting.

■ DOWNY MILDEW

Cause

The oomycete *Hyaloperonospora parasitica*.

Symptoms

Young seedlings are more usually seriously affected by downy mildew. Irregular, yellow patches develop on the cotyledons or young leaves. A white fungal growth develops on the undersides of leaves and may progress to the upper surface during cool moist conditions. Conidiophores bearing spores on branches may be visible with the aid of a hand lens. Severely affected seedlings may die.

Fig 7.16 Downy mildew of brassica seedlings. Note the grey-white spore clusters.

Older leaves develop chlorotic areas and small water-soaked necrotic areas which give a flecked or speckled appearance. In moist weather, spots enlarge to form large patches, but with a return to dry weather, these dry out and die. Affected cauliflower heads may show a dark brown discolouration. Leaves of seedlings may be completely blighted.

Blue root disease of Red Globe radish varieties causes blue to black lesions, up to 15 mm wide, on the swollen roots. Root russeting cracking and distortion may occur after the early infection. Downy mildew lesions on the edible root usually occur on the upper half and are independent of emerging rootlets. This distinguishes the disease from black root caused by *Aphanomyces raphani*.

Source of infection and spread

The pathogen survives on volunteer brassicas and weeds such as shepherd's purse (*Capsella bursa-pastoris*) and hedge mustard (*Sisymbrium officinale*). The mildew produced on

Fig 7.17 Downy mildew on cabbage showing leaf speckling.

Fig 7.18 Dark brown discolouration of cauliflower heads affected by downy mildew.

the underside of the leaves contains large numbers of spores that spread in wind and water. Plant infection occurs through the leaf cuticle or stomata. Cool, humid, weather favours rapid development of the disease. Downy mildew is most serious in the seedbed or in seedling plant houses, where crowding and extended periods of leaf wetness provide ideal conditions for the disease to develop.

Importance
Downy mildew is often a serious disease in seedlings, but is usually a minor disease after transplanting.

Management
- Improve air movement around young plants and irrigate early in the day to allow plants to dry.

- Spray with the recommended chemicals, particularly in seedling production areas. Protectant chemicals should be applied frequently (every seven days), before symptoms of the disease develop. Good coverage of the underside of leaves is essential.

- Keep seedling production areas free from susceptible weeds.

- Rotate brassicas with non-brassica species and control brassica weed species that may provide alternative hosts for the pathogen.

Fig 7.19 Downy mildew in radish.

■ LIGHT LEAF SPOT

Cause
The fungus *Cylindrosporium concentricum*.

Symptoms
Silvery lesions appear on the upper leaf surface and become pale and diffuse with age. Concentric rings of pale pink to white dots appear on fully expanded leaves. The pathogen

Fig 7.20 Petiole symptoms of light leaf spot.

Fig 7.21 Close view of light leaf spot symptoms.

Fig 7.22 Light leaf spot symptoms.

Fig 7.23 Ringspot symptoms showing circular, dark spots surrounded by a yellow halo.

grows under the leaf cuticle and disrupts the waxy surface as it sporulates, causing the cuticle to lift and bleached areas appear, giving the name light leaf spot to the disease. Lesions are often surrounded by numerous dark fruiting bodies around which white drops of spores form in concentric rings.

Dark brown lesions may also develop on petioles and cauliflower curds.

Source of infection and spread

The fungus survives on crop debris and in canola and brassica vegetable crops. Fungal spores are dispersed by wind, water splash and wind driven rain. The disease is favoured by cool, wet conditions. Infection requires free water and at least six hours of leaf wetness at 16°C or longer periods at lower temperatures.

Importance

Light leaf spot can reduce the quality of cauliflower, Brussels sprouts and other vegetable brassicas when grown under cool, wet conditions. Canola is highly susceptible.

Management

• Destroy crop residues to reduce carry over of the pathogen.

• Rotate brassicas with other crops.

• Avoid planting vegetable brassicas near canola crops.

■ RINGSPOT

Cause

The fungus *Mycosphaerella brassicicola*.

Symptoms

Small, dark, circular spots develop on leaves, enlarging up to 20 mm in diameter and becoming light brown to grey

towards the centre. A yellow halo often surrounds spots. Small, black fruiting bodies form on the surface of spots, generally in concentric circles.

Source of infection and spread

The fungus survives from season to season on undecomposed crop residues in the soil or in old brassica crops. Ascospores are forcibly ejected from the fruiting bodies and disperse in the wind. Cool, wet weather favours infection and disease development. The fungus may also be seed-borne.

Importance

Ringspot can be serious on Brussels sprouts, cauliflower and cabbage grown in cool, wet production areas.

Management

• Plough in diseased crops immediately after harvesting. Do not replant until all residues have completely decomposed.

• Rotate brassica crops with unrelated species.

Fig 7.24 Close-up view of ringspot lesions.

- Apply the recommended fungicides. Base spray programs on regular applications of protectant fungicides, with targeted applications of systemic products following cool, moist weather.

■ STEM CANKER

Cause
Several pathogens are associated with the disease: *Rhizoctonia, Phoma, Fusarium, Sclerotinia* and *Pythium*. *Rhizoctonia solani* (several anastomosis groups) and *Leptosphaeria maculans* appear to be the dominant pathogens.

Symptoms
The symptoms range from superficial scurfing and stem russetting to discrete stem lesions or a complete stem rot and collapse.

Source of infection and spread
The pathogens are soil-borne also surviving in crop residues, and, possibly, alternative crop and weed hosts.

Importance
Stem canker occurs to a varying extent in most brassica production areas of Australia. Cauliflower is most affected but the disease also occurrs in Brussels sprouts, red and green cabbage and broccoli. Losses of up to 80% of plants have been reported in cauliflower crops.

Management
- Rotate brassica crops with unrelated species.

- Plant resistant varieties or less susceptible varieties.

Fig 7.26 Advanced symptoms of stem canker.

■ WHITE BLISTER (WHITE RUST)

Cause
The oomycete *Albugo candida*.

Symptoms
Small, raised white or cream coloured blisters (sori) form on the undersides of leaves and can also develop on stems, heads and flower stalks. The blisters contain spores that develop under the plant epidermis. The epidermis

Fig 7.27 White blister on broccoli heads, showing stag head symptoms (top).

Fig 7.25 Early symptoms of stem canker.

Fig 7.28 Detail of white blister.

The presence of blisters on broccoli and cauliflower heads reduces quality or renders them unmarketable. The quality of Chinese cabbage and radish can be severely reduced by white rust infection.

The pathogen occurs as a complex of races based on the ability to infect various brassica species. For example, the races infecting several brassica weed species differ from those infecting broccoli and cauliflower, while races from radish are unlikely to infect other brassica species.

For many years, white blister has caused economic losses in Australian crops of rocket, Chinese cabbage and radish. Since 2002, the disease has become a major problem in broccoli and, to a lesser extent, in cauliflower and Brussels

eventually breaks, releasing white powdery spores that disperse in wind. As blisters on the undersides of leaves age, light green to yellow spots develop on the upper leaf surface and these may also produce spores. Blisters may continue to grow to 2–3 cm in diameter and leaves may become distorted and eventually wither and die.

Systemic infection of plants may also occur and causes twisting and growth deformities in stems, leaves and flowers. The twisted, deformed growth of flowering heads is called 'staghead' because of its spiny, antler-like appearance.

The downy mildew pathogen is frequently associated with white rust and the pathogen is often found on white rust affected plant parts, including stagheads.

Source of infection and spread

The pathogen survives as resistant oospores in crop residues, on mature infected crops and in alternative brassica hosts. The oospores germinate, releasing zoospores that are splashed or swim in moisture to infect susceptible hosts and raised blisters are formed. Air-borne spores produced in the blisters spread throughout the crop in wind. Zoospores, also produced on the surface of the host plant, directly penetrate the epidermis. Infection and disease development occurs when leaves are moist for at least two hours and temperatures are in the range of 13–25°C. Disease severity increases as the duration of leaf wetness increases.

Importance

White blister occurs almost wherever brassica crops are grown. Brassica types affected include broccoli, Brussels sprouts, cauliflower, radish, Chinese cabbage, rocket, turnip and cress.

Fig 7.29 White blister pustules on cabbage, showing systemic leaf infection (bottom).

sprouts grown in southern Australian production areas. The disease has been found in Queensland broccoli crops since 2005.

Management

- Use high quality seed and plant disease-free transplants.

- Ensure that all crop residues are thoroughly decomposed before sowing.

- Rotate crops. Avoid consecutive plantings of brassicas and rotate with non-brassica crops.

- Avoid long periods of leaf wetness as spores need to water to germinate.

- Maintain good air movement within crops through row orientation, increased plant spacings and avoiding areas prone to heavy fog and mist.

- Grow varieties tolerant or resistant to white rust. No varieties of broccoli with resistance to white blister are currently available.

- Apply recommended fungicides. Apply these before the disease becomes serious and use them as part of a management strategy for fungicide resistance. White blister has the capacity to develop resistance to fungicides very quickly, particularly where systemic fungicides are overused and/or not combined, or alternated with protectant chemicals.

■ YELLOWS (FUSARIUM WILT)

Cause

The fungus *Fusarium oxysporum* f.sp. *conglutinans*.

Fig 7.30 Brussels sprouts plants affected by yellows (Fusarium wilt).

Symptoms

Affected plants lose vigour and the lower leaves turn yellow. In young plants, leaves tend to bend sideways, with one side retarded in growth and pale yellow in colour. Symptoms may show on one side of the plant only.

If the stem is cut across near ground level, a brown discolouration of the water conducting tissues just beneath the bark can be seen. The one-sided development of leaves results from more advanced browning on one side of the stem.

Source of infection and spread

The fungus may survive for long periods in the soil as resistant chlamydospores. The fungus infects the plant through the roots and grows into the water-conducting tissues. Warm weather (24–29°C) favours the disease. The fungus spreads in contaminated soil and water and in infected transplants. The fungus may also infect brassica weed species.

Importance

Cabbage is the major crop affected by yellows but cauliflower, broccoli, radish and other brassicas are also susceptible. Resistant varieties usually provide good control.

Management

- Plant resistant varieties.

- Prevent the movement of contaminated soil to clean fields. Decontaminate machinery by washing with the detergent-based degreaser Farmcleanse® or a quaternary ammonium based sanitiser.

VIRUSES

■ TURNIP MOSAIC

Cause

Turnip mosaic virus (Potyvirus).

Symptoms

Mosaic and yellow ringspots develop on leaves. The ringspots may develop into yellow or brownish spots surrounded by irregular necrotic rings on older leaves.

Necrotic streaks and flecks may also develop on plants, which become deformed and stunted.

Fig 7.31 *Turnip mosaic virus* on cabbage plants.

Fig 7.32 *Turnip mosaic virus* symptoms on cabbage.

Internal dead or necrotic areas may develop in stored cabbage, and may be seen only when the cabbage is cut.

Source of infection and spread
The virus has a wide host range among cultivated and brassica weed species. It also infects crop and weed hosts in other plant families, for example lettuce and rhubarb. The virus is spread by aphids in a non-persistent manner and is acquired and transmitted during short feeding periods.

Importance
Turnip mosaic is the most common virus found in brassica crops and can cause significant crop losses.

Management
- Control weeds around the crop area.

- Plant healthy transplants.

- Avoid growing cabbages in very cold weather if the disease has been prevalent.

- Use tolerant or resistant varieties.

■ BEET WESTERN YELLOWS

Cause
Beet western yellows virus (BWYV; Polerovirus).

Symptoms
The virus causes lower leaves to redden and plants to become stunted. The symptoms are not diagnostic of viral infection and can be easily confused with those caused by nutrient imbalance, herbicide damage and other stress factors.

Source of infection and spread
The virus is spread by aphids in a persistent manner, which requires aphids to feed for several hours to both acquire the virus from an infected plant and then transmit the virus to another plant.

BWYV has a wide natural host range including cultivated and weed species of brassicas, lettuce and beet. Wild radish (*Raphanus raphanistrum*) is a common host of the virus in Western Australia and the southern eastern States.

Importance
The virus is not usually an important cause of loss in vegetable brassicas in Australia.

Management

- Control weeds, particularly brassica weeds, in and around crops, as these are hosts of both the virus and aphid vectors.

- Avoid planting new crops adjacent to older plantings.

- Apply registered insecticides as required to control aphids. The long feeding periods required for transmission allow insecticides to kill aphids before the transmission process is complete.

■ CAULIFLOWER MOSAIC VIRUS

Cause
Cauliflower mosaic virus (CaMV; Caulimovirus).

Symptoms
Leaves develop mosaic and vein clearing. Infected plants may also remain symptomless, particularly when grown under warm temperatures. Severely affected plants appear stunted, with smaller, poorer quality heads. Symptoms are generally more severe in turnip and Chinese cabbage.

Source of infection and spread
The virus has a wide host range among brassica species, including several common brassica weeds such as wild turnip.

The virus is aphid transmitted with the cabbage aphid (*Brevicoryne brassicae*) and the green peach aphid (*Myzus persicae*) often been important vector species. Aphids can acquire and transmit the virus after feeding periods of only several minutes.

Importance
The virus is present in most temperate areas where brassica crops are grown and causes sporadic important crop losses.

Management
- Plant healthy transplants.

- Control brassica weeds in and around crops.

- Avoid planting new crops adjacent to older plantings.

■ FURTHER INFORMATION

Donald C (2005). *Clubroot factsheets.* Department of Primary Industries, Victoria.

Donald C, Endersby N, Ridland P, Porter I, Lawrence J, Ransom L (2000). *Field guide to pests, diseases and disorders of vegetable brassicas.* Department of Natural Resources and Environment, Victoria.

Minchinton E (2005). *White blister control notes.* Department of Primary Industries, Victoria.

Rimmer RS, Shattuck VI & Buchwaldt L (Eds) (2007). *Compendium of brassica diseases.* APS Press, St Paul, Minnesota.

Capsicum or pepper is consumed as both a fresh vegetable, and a dehydrated spice. Most cultivated capsicums are classified as *Capsicum annuum* and belong to family Solanaceae. Capsicum originated in the tropical Americas and a great diversity of fruit shapes, sizes, colours, fruit pungency and growth habit occurs within the genus.

The majority of capsicums grown in Australia are sweet bell types, but sweet long fruited and hot chilli varieties have a market niche.

Most crops are produced from transplants grown in commercial nurseries. Capsicums are grown in all States with Queensland producing most of the crop in Bundaberg and north Queensland during the cooler months.

Bacterial leaf spot, soil-borne fungal wilts and tospovirus infection are the more common diseases affecting production.

BACTERIA

■ BACTERIAL SPOT

Cause

The bacterium *Xanthomonas campestris* pv. *vesicatoria*. Numerous races of this pathogen have been identified, including races pathogenic only to either tomato or capsicum or races pathogenic to both hosts.

Fig 8.2 Bacterial spot causing yellowing and defoliation.

Symptoms

On leaves, small, angular, water-soaked spots develop that later turn brown and seldom exceed 5 mm in diameter. In favourable conditions, the spots often coalesce and cause leaf blight. Severely diseased leaves, particularly lower leaves, fall from the plant.

Fruit occasionally develop dark brown to black scabs, often concentrated on the shoulders.

Source of infection and spread

The bacterium is commonly introduced in seed. Once established as leaf spots, the bacteria spread during wet, windy weather and in overhead irrigation. Warm (24–30°C) and humid weather favours the disease.

Fig 8.1 Early symptoms of bacterial spot on capsicum leaves.

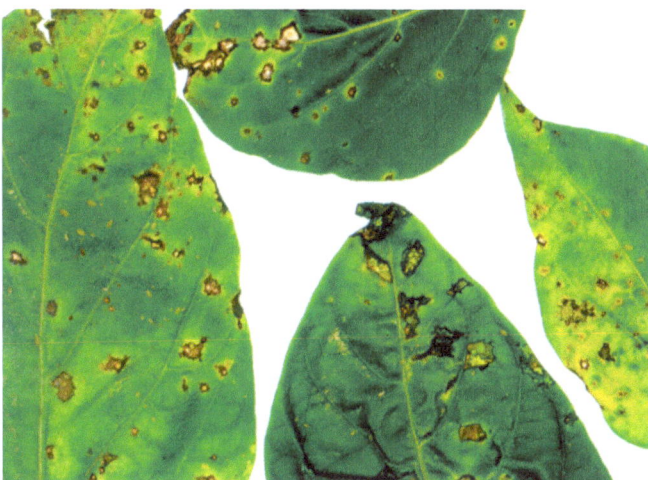

Fig 8.3 Examples of the range of leaf symptoms caused by bacterial spot.

Bacterial spot also affects tomatoes and may survive on undecomposed tomato or capsicum residues in the soil. Weed hosts such as black nightshade (*Solanum nigrum*) are another source of infection.

Importance
Bacterial spot is a sporadic disease that may be serious, particularly in wet, humid weather.

Management
- Use healthy seed to produce seedling plants.

- Use resistant varieties.

- Apply preventative copper sprays in the nursery and field, particularly if wet or showery weather is expected. Copper should be tank-mixed with mancozeb to improve spray efficacy.

- Twin-jet nozzles should be used to apply foliar sprays in order to maximise plant coverage.

- Destroy old capsicum and tomato crops promptly after harvest.

- Rotate capsicum crops with crops other than tomato.

■ BACTERIAL WILT

Refer to the description of bacterial wilt in the chapter on Tomato diseases.

FUNGI

■ ALTERNARIA FRUIT ROT

Cause
The fungus *Alternaria* spp.

Symptoms
Dark, sunken spots up to 30 mm wide, develop on fruit and may become covered with a brown to black velvety mould. The symptoms generally develop following injury, e.g. blossom end rot or sunscald.

Source of infection and spread
The fungus survives on crop debris, and spores are spread by wind. Infection occurs through injured tissue. Wet weather favours the disease.

Importance
The disease usually causes only minor losses of fruit.

Fig 8.4 Alternaria fruit rot.

Management
- Spray with the recommended fungicides, particularly during wet weather.

■ ANTHRACNOSE

Cause
The fungi *Colletotrichum* spp.

Symptoms
Symptoms usually appear after fruit begin to ripen. Small, indefinite, circular, slightly sunken, dark yellow spots develop on the fruit surface. The spots may enlarge and coalesce when conditions favour the disease. Black acervuli (fruiting bodies) may develop from the darkened centre of the spot, arranged in concentric rings within the fruit spot. Black hairs (setae)

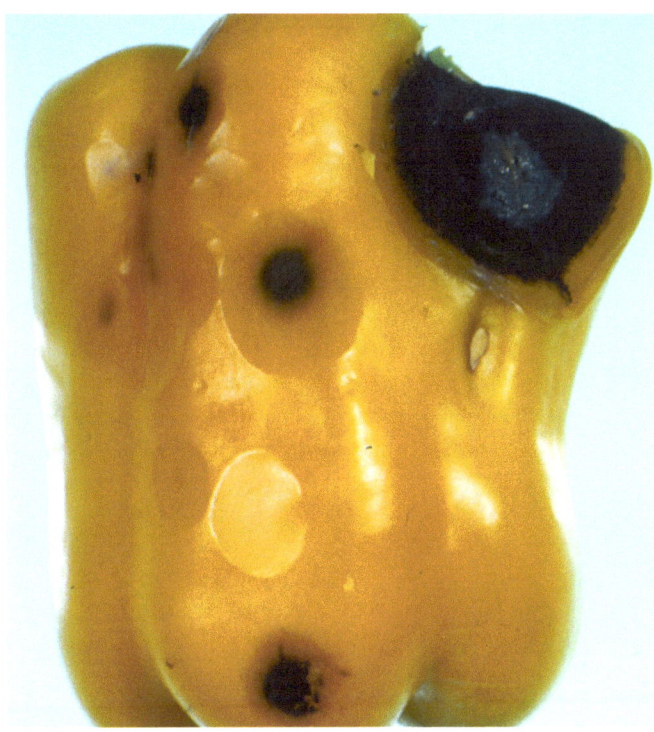

Fig 8.5 Anthracnose on capsicum fruit.

may develop on spots, depending on the *Colletotrichum* species involved, causing the spot to darken in colour. Pink spore masses may form on the spots during moist weather. The rot can extend into the seed cavity and infect the seed.

Source of infection and spread

The fungi survive on crop residues and on or in capsicum seed. Large numbers of spores produced on fruit spots spread to healthy fruit during wet, windy weather. Although infection may occur at any stage of fruit development, symptoms do not show until ripening. Fruit in contact with the soil can become infected by soil-borne inoculum.

Importance

Although the pathogen is widespread in capsicum growing regions, losses from anthracnose are usually minor. The disease occurs most often in chilli crops.

Fig 8.6 Anthracnose on chilli fruit.

Management

- Spray preventatively with recommended fungicides.

- Avoid sprinkler irrigation as water splash spreads fungal spores.

- Plant disease-free seed. Seed known to be affected should not be planted. Seed treated with hot water (52°C for 30 minutes) can be used to reduce infection, but care should be taken with hot water treatment as germination may be affected adversely.

- In fields with a history of the disease, a three-year rotation to non-solanaceous crops will reduce inoculum levels.

- Following harvest, plough in or remove crop residues promptly in infected fields.

■ BASAL STEM AND ROOT ROT

Cause

The fungus *Fusarium oxysporum*.

Fig 8.7 Basal stem and root rot caused by *Fusarium oxysporum*.

Fig 8.8 Stem lesions caused by *Fusarium oxysporum.*

Symptoms

A dark brown rot develops in the roots and crown. Extensive discolouration of the roots and crowns occurs and brown lesions may develop on the external tissues of the lower stems. When the root and crown tissues are sufficiently compromised the entire plant wilts.

Source of infection and spread

This disease has not been well characterised in Australia. Disease outbreaks most commonly occur in nurseries or soon after seedlings are transplanted to the field, indicating that seed-borne infection is likely.

Importance

This disease occurs infrequently in Queensland, however plant losses may be significant, particularly in nursery situations.

Management

• Start with clean planting material. Infected seedlings should not be transplanted to the field.

■ CERCOSPORA LEAF SPOT

Cause

The fungus *Cercospora capsici*.

Symptoms

Small, round water-soaked lesions develop on leaves, petioles and stems. The lesions enlarge and have light

Fig 8.9 Cercospora leaf spot.

brown centres with dark brown-red margins. As lesions expand, an outer water-soaked area and dark ring may form beyond the original lesion margin, so that the lesion centre becomes surrounded by concentric rings. With age, the lesion centres dry out and crack.

Source of infection and spread

Warm, wet weather favours the disease. Infection from germinating fungal spores occurs via penetration of leaf stomata by fungal hyphae. Spores spread in wind, rain, irrigation or via mechanical means. The fungus is likely to carry over to new crops on infected crop debris.

Importance

Cercospora leaf spot is a common, but usually minor disease of capsicum in tropical and subtropical areas.

Management

• Specific controls are not usually required.

PHYTOPHTHORA BLIGHT – BIOSECURITY THREAT

Cause

The oomycete *Phytophthora capsici*.

Symptoms

The pathogen causes a crown and root rot as well as aerial symptoms. A dark rot develops in the roots and crown tissues that are below the soil line, and crown lesions can develop several inches up the stem. Root infection causes the plant to wilt. Aerial symptoms may include dark brown lesions on upper stems, as well as circular grey-brown water-soaked leaf lesions. Fruit rot can also occur. Nearly complete loss of plants can result if fields are infested heavily.

Source of infection and spread

Root and crown symptoms arise from direct infection via soil-borne spores. The aerial phase of the disease arises when the spores are splash-dispersed up into the canopy from soil, or from the above-ground portion of crown lesions. The pathogen persists in soil as hard-walled resilient oospores. It also produces sporangia that give rise to motile zoospores that swim through soil moisture to invade susceptible plant roots. The disease can also spread between healthy and infected plants via root-to-root contact. Spread of the spores in splashing water and surface water allows the pathogen to disseminate rapidly. The pathogen also spreads between aerial plant parts as water splashes from infected leaves, stems and fruit. The pathogen has a wide host-range across vegetable crops.

Fig 8.11 Stem lesion caused by *Phytophthora*.

Fig 8.10 Phytophthora blight on capsicum fruit.

Fig 8.12 Phytophthora blight symptoms on stems and leaves.

Importance

Phytophthora blight is a widespread and devastating disease of capsicum worldwide. It is the most severe capsicum disease in the United States and causes heavy losses in capsicum and chilli crops throughout Asia. Although *P. capsici* has been reported on other hosts in Australia, disease associated with this pathogen has never been reported on capsicum and chilli in Australia.

What to do if you suspect Phytopthora blight

This pathogen is a biosecurity risk to Australia. Any suspected affected plants should be reported to the nearest Department of Primary Industries or the Plant Health Australia hotline (1800 084 881).

■ POWDERY MILDEW

Cause

The fungus *Leveillula taurica* (anamorph *Oidiopsis sicula*).

Symptoms

Light green to yellow spots or blotches develop on older leaves. A white, powdery fungal growth (mycelium) occurs on the undersides of these leaves. The spots coalesce to cover much of the leaf, which turns yellow, curls upwards and falls from the plant. Under favourable conditions, extensive defoliation occurs, exposing maturing fruit to sunburn.

Fig 8.15 Defoliation in plants severely affected with powdery mildew.

Source of infection and spread

The fungus is an obligate parasite and will only survive and reproduce on living plants. Host plants include all capsicum (pepper) types, tomato and several weed species. Fungal spores disperse in wind and warm, dry weather favours the disease. If favourable environmental conditions persist throughout the growing season, inoculum pressure will gradually increase. The disease tends to be most severe at the end of the cropping period.

Fig 8.13 Advanced powdery mildew on capsicum showing chlorotic blotches and leaf wilting.

Importance

Powdery mildew is a common and often serious disease in warm, dry areas.

Management

- Varieties differ in resistance to powdery mildew. Select a less susceptible variety if conditions are likely to favour the disease.

- Apply silicon sprays at the beginning of the spray program and again at fruit set. Silicon strengthens cell walls and activates the plant's defence mechanisms against the fungus.

- Apply fungicides as soon as symptoms of powdery mildew are seen. Alternate between systemic and protectant fungicides. Good spray coverage is essential.

Fig 8.14 Symptoms of leaf curl and leaf blotching caused by powdery mildew.

Fig 8.16 Stem rot caused by *Sclerotium rolfsii*.

■ STEM ROT

Cause
Sclerotium rolfsii.

Refer to the chapter on Common diseases of vegetable crops.

■ SUDDEN WILT OR MATURE PLANT COLLAPSE

Cause
An interaction between soil-borne *Pythium* species and high soil temperatures.

Symptoms
Plants are healthy until fruit set begins. Early symptoms are plants wilting during the warmer parts of the day and becoming stunted, leaves yellowing and some fruit shrivelling. Plants then wilt completely, lose their leaves, and set only small fruit. Affected plants often die. When above-ground symptoms are first noticed, the root system

Fig 8.17 Collapsed plants affected by sudden wilt.

is generally severely rotted. Rotting occurs first on the small roots and progresses to the larger roots, destroying most of the root system.

Source of infection and spread
Although several pathogens have been isolated from the roots of plants affected by sudden wilt, two species of *Pythium* (*Pythium myriotylum* and *P. aphanidermatum*) are consistently associated with the disease and destroy the root system of capsicum plants. This destruction is more severe and rapid under high soil temperatures, resulting in plants wilting rapidly once fruit set begins. The plant is then unable to obtain the large quantity of water required during the fruit set period.

The *Pythium* species occur widely in most crop production areas and infect the root systems of many crop and weed species.

High soil temperatures and waterlogged, compacted soils with low oxygen levels favour the activity of *Pythium*.

Importance
Sudden wilt can cause serious crop losses during hot, dry weather or when crops are grown in plastic mulch, which results in soil temperatures greater than 35°C. In north Queensland, losses are most likely in crops planted in early autumn when temperatures are high, while losses in southern Queensland usually occur when crops are planted into black plastic mulch to increase soil temperatures and enhance the early growth of winter-planted crops.

Management
• Maintain soil structure using green manure crops and other measures. Compacted soils with poor structure, poor drainage and low aeration favour soil-borne pathogens.

- When preparing hills for beds with plastic mulch, ensure soil depth and drainage is adequate.

- Use white, reflective plastic mulch instead of black mulch during hot weather.

- Do not plant oversized, root-bound seedlings as deep planting predisposes plants to sudden wilt.

- Use irrigation scheduling to ensure plants are neither over-watered (which causes waterlogging and low oxygen levels), nor stressed through under-watering.

VIRUSES

■ *CAPSICUM CHLOROSIS VIRUS*

Cause
Capsicum chlorosis virus (CaCV; Tospovirus).

Symptoms
Infected plants are often stunted, and leaves show chlorotic blotches, mottling and chlorotic ring and line patterns. Young leaves develop chlorosis between the leaf veins and along the leaf margins. These leaves often become narrow and curled.

Young leaves of plants infected by CaCV seldom develop the bright chlorotic blotches and ringspots typical of *Tomato spotted wilt virus* infection in capsicums.

Fruit is distorted, generally reduced in size and has blotches or dark spots on the skin, giving the fruit a scarred appearance.

Fig 8.19 Chilli fruit infected with *Capsicum chlorosis virus*.

Source of infection and spread
Refer to the chapter on Common diseases of vegetable crops.

Importance
CaCV is common in Queensland capsicum crops south to and including Bundaberg where it causes sporadic serious epidemics. The virus is also found in southern Queensland, New South Wales and in the Kununurra area of Western Australia.

Management
- Use resistant varieties.

- Maintain a high standard of farm hygiene by removing old crops promptly and controlling weeds in and around crops that are hosts of both the virus and thrips carriers.

Fig 8.18 Leaf symptoms of *Capsicum chlorosis virus*.

Fig 8.20 Capsicum fruit affected by *Capsicum chlorosis virus*.

- Plant healthy transplants.
- Monitor thrips populations and control as necessary.

■ CAPSICUM YELLOW VEIN

Cause
Ranunculus white mottle virus (Ophiovirus).

Symptoms
Affected plants develop conspicuous yellowing in leaf veins. Symptoms first appear on the youngest leaves with most leaves affected as the disease progresses. Fruit may develop a slight mottle and be reduced in size. Symptoms often fade as the plant matures.

Source of infection and spread
The source of the virus and the means of transmission are unknown at present. Viruses in the Ophiovirus group spread with the soil-borne fungus *Olpidium*. Infection occurs when pathogen zoospores enter the plant roots.

Hosts of *Ranunculus white mottle virus* in other countries are freesia and anemone.

Importance
The disease has been prevalent in glasshouse grown capsicum crops on the North Adelaide plain for several years. The viral cause of the disease was determined in 2008. The yields from affected plants do not appear seriously reduced.

Management
No specific methods are available.

■ *CUCUMBER MOSAIC VIRUS*

Cause
Cucumber mosaic virus (CMV; Cucumovirus).

Symptoms
Symptoms vary considerably depending on the virus strain, capsicum variety and weather. Symptoms include mosaic or mottling patterns and chlorotic or necrotic flecks on leaves. Fruit may be deformed with ringspots or chlorotic blotches.

Source of infection and spread
The virus has a very wide host range among crops, ornamentals and weeds. CMV is spread by many species of aphids and only very short feeding times are required for transmission.

Importance
CMV causes sporadic losses in Australian capsicum crops.

Management
- Control alternative weed and crop hosts in and around capsicum crops.
- Avoid planting young crops adjacent to older crops that may harbour virus inoculum and aphid carriers.

Fig 8.21 Capsicum yellow vein disease.

Fig 8.22 Leaf symptoms of *Cucumber mosaic virus* infection.

■ *PEPPER MILD MOTTLE VIRUS*

Cause
Pepper mild mottle virus (PMMoV; Tobamovirus).

Symptoms
Affected plants are often stunted with mild leaf-yellowing. The most obvious symptoms are small and malformed fruit with necrotic spots or mottle patterns on the skin. The Tobamoviruses Tobacco mosaic and Tomato mosaic also infect capsicums, causing similar symptoms.

Source of infection and spread
The virus is seed-borne and can be carried both on the surface of the seed and as an internal seed infection. The virus can survive in infected crop residues in the soil for several months. Handling and touching plants can spread the virus from plant to plant. It survives on tools, clothing, stakes and other equipment. PMMoV has no known insect carrier. The natural hosts are confined to *Capsicum* species.

Importance
PMMoV occurs in most countries where capsicums are grown and a high level of infection reduces the yield of marketable fruit substantially. The disease is often common in crops grown in glasshouses or plastic tunnels.

Management
* Treat seed with tri-sodium phosphate to remove seed-borne virus.
* Avoid excessive handling of plants.
* Resistant varieties are available.

Fig 8.23 Fruit infected with *Pepper mild mottle virus*.

Fig 8.24 *Potato virus Y* on a capsicum leaf. Note the dark green vein banding.

■ MOSAIC

Cause
Potato virus Y (PVY; Potyvirus).

Symptoms
Leaves are puckered with narrow, discontinuous bands of dark green tissue along the main veins. This contrasts with areas of abnormal, yellow-green tissue between the veins. Affected plants appear stunted and yields reduced. Fruit may be malformed and develop mosaic patterns on the skin.

Source of infection and spread
Mosaic is caused by the same strain of *Potato virus Y* that causes leaf shrivel in tomato and vein necrosis in tobacco. It is spread into and within capsicum crops by aphids in a non-persistent manner. Several weeds of the potato family (Solanaceae) are also hosts, e.g. apple of Peru (*Nicandra physalodes*) and nightshades (*Solanum* spp.).

Importance
Mosaic can be serious in susceptible varieties.

Management
* Grow resistant varieties.
* Destroy old infected capsicum and tomato crops. Ratooning capsicum favours the disease.
* Control weeds in and around crops as these may provide hosts for both the virus and aphid vectors.
* Avoid planting new crops adjacent to mature crops of capsicum and tomato.

■ SPOTTED WILT

Cause
Tomato spotted wilt virus (TSWV; Tospovirus).

Fig 8.25 Foliar symptoms of *Tomato spotted wilt virus*.

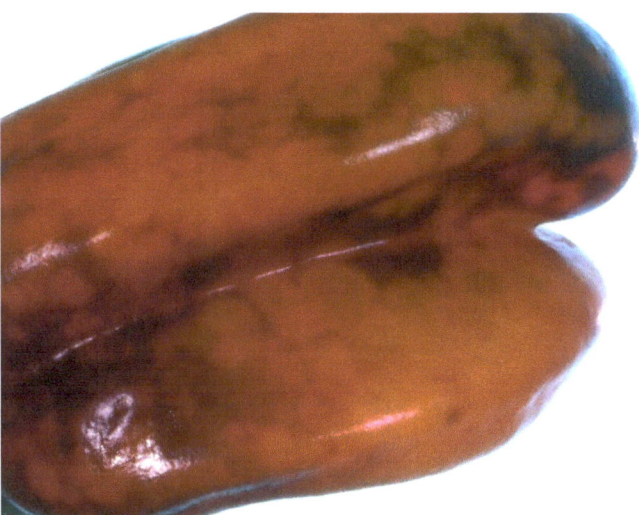

Fig 8.27 Fruit infected with *Tomato spotted wilt virus*.

Symptoms

Infected plants are usually stunted and leaves develop chlorotic blotching, mottling and chlorotic ring and line patterns. Fruit is distorted, usually reduced in size and has blotches or dark spots on the skin, giving a scarred appearance. Plants infected as the fruit mature may only develop symptoms on fruit.

Importance

TSWV frequently causes considerable losses in capsicum crops in Australia. Refer to the chapter on Common diseases of vegetable crops.

Management

Refer to the section on *Tomato spotted wilt virus* in Common diseases of vegetable crops.

■ FURTHER INFORMATION

Bosland PW, Votava EJ (2000). *Peppers: vegetable and spice capsicums.* CABI Publishing, Wallingford.

Meurant N, Wright R, Olsen J, Fullelove G and Lovatt J (1999). *Capsicum and chilli information kit.* Agrilink series. Department of Primary Industries and Fisheries, Brisbane, Queensland.

Pernezny K, Roberts PD, Murphy JF and Goldberg NP (Eds) (2003). *Compendium of pepper diseases.* APS Press, St Paul, Minnesota.

Stirling GR, Eden LM (2008). The impact of organic amendments, mulching and tillage on plant nutrition, Pythium root rot, root-knot nematode and other pests and diseases of capsicum in a subtropical environment, and implications for the development of a more sustainable vegetable farming system. *Australasian Plant Pathology* **37**, 123–131.

Zitter TA (1989). *Phytophthora blight of cucurbits, pepper, tomato and eggplant.* Cornell University, Department of Plant Pathology, Ithaca, NY. http://vegetablemdonline.ppath.cornell.edu/factsheets/Cucurbit_Phytoph.htm

Fig 8.26 Examples of the range of leaf symptoms caused by *Tomato spotted wilt virus*.

Carrot (*Daucus carota*) is the most important and widely grown member of the family Apiaceae. Carrots are grown for the edible taproot, which contains high levels of carotene, the precursor of vitamin A, which is essential for human health and nutrition.

The major centre of genetic diversity and most likely centre of origin for carrots is the Afghanistan and Turkestan region.

Carrots are grown on a wide range of soil types, although sandy soils are preferred because they produce better root quality and make harvesting relatively easy. The crop grows from seeds sown into beds. Crops mature between 14 and 24 weeks, depending largely on temperature.

Carrots are susceptible to a range of diseases and the risk of economic losses from disease increases with short rotations, overlapping crops and wet weather.

BACTERIA

■ BACTERIAL LEAF BLIGHT

Cause
The bacterium *Xanthomonas campestris* pv *carotae*.

Symptoms
The first symptoms on leaves are small, yellow, angular spots that quickly expand into irregular, brown water-soaked spots

Fig 9.2 Bacterial leaf blight on carrot leaf.

surrounded by a yellow halo. The centres of the spots become dry and brittle. The spots are more common on the margins of leaves, especially at the V-shaped junction of the leaflet lobes. Affected leaflets may be curled and distorted. The bacterium may also cause a gummy exudate and the petioles to turn brown. The symptoms of bacterial leaf blight can be similar to those caused by other foliar pathogens such as the fungi *Alternaria* and *Cercospora*.

Source of infection and spread
Contaminated seed is the pathogen's most important means of long-distance dissemination. The bacteria are dispersed in water splash from rain or overhead irrigation and can also spread with insect activity and machinery moving through the crop. The pathogen can also survive in the soil on infested crop residues. Disease development is favoured by warm (25–30°C), wet weather, and by the extensive use of overhead irrigation.

Management
- Plant high quality, disease-free seed. Grow seed crops in dry areas without overhead irrigation to reduce the risk of seed-borne infection. Infected seed can be hot water treated (50°C for 20 minutes) to reduce infection levels.

- The pathogen has a limited host range so rotate carrots with other crops in a two or three year cycle to prevent disease carry-over and limit the accumulation of infested crop residues.

Fig 9.1 Bacterial leaf blight on carrot stem and leaf.

FUNGI

■ BLACK ROOT ROT

Cause
The fungi *Thielaviopsis basicola* (*Chalara elegans*) and *Chalaropsis thielavioides*.

Symptoms
This disease is predominantly a post-harvest problem that may develop when washed roots are stored incorrectly. Superficial dark grey to black fungal growth develops on the surface of affected roots. Infected tissues turn black and have irregular margins. Tissue below the surface remains firm unless secondary bacterial infection occurs.

Source of infection and spread
Both fungi are common soil-borne organisms and survive by means of thick-walled chlamydospores. Carrot roots are likely to be infected by the fungi prior to harvest or through wounds and abrasions during harvest, washing and grading. Disease symptoms can develop rapidly in carrots held after harvest without rapid cooling.

Importance
Potentially, the disease can cause extensive losses when carrots are not correctly stored after harvest. Washed carrots packed in plastic are particularly vulnerable due to high humidity in the packages.

Management
• Minimise damage to carrots during harvest, washing and grading.

• Rapidly cool carrots after harvest and hold at low temperature (5°C).

■ CARROT BLACK ROT

Cause
The fungus *Alternaria radicina*.

Symptoms
The pathogen can cause poor seedling establishment, damping-off, leaf and crown infection, flower blight and a storage rot.

Carrots that survive early infection frequently develop a black ring of decay around the top of the stem. Senescing leaves are often infected first, followed by a crown infection that may cause a rot in the upper portion of the storage root.

The fungus may also cause small brown leaf spots with a chlorotic halo, later enlarging to form black lesions.

Source of infection and spread
The fungus is commonly seed-borne, including seed lots that have been treated with fungicides. The fungus is found both on the surface of the seed and in the seed pericarp and testa. *Alternaria radicina* also survives in crop residues and as microsclerotia that can survive in soil for several years.

Warm, humid, wet weather favours disease development.

Importance
The pathogen is seed-borne and occurs worldwide in most carrot production areas. Infected seed-lots can lead to poorly established crops and crown infections. Other

Fig 9.3 Black root rot.

Fig 9.4 Carrot black rot caused by *Alternaria radicinia*.

members of the Apiaceae are also susceptible, including parsnip, parsley and celery.

Management

- Use seed free of the pathogen or with low levels of contamination.
- Treat seed with hot water or fungicides.
- Rotate carrots with unrelated crops.

■ CAVITY SPOT AND FORKING

Cause

Cavity spot is caused by the oomycetes *Pythium sulcatum* and *Pythium violae*. The former occurs in all carrot-growing regions of Australia, while *P. violae* is restricted to irrigated production areas along the Murray River in Victoria and South Australia.

Symptoms

Symptoms usually develop about four weeks before carrots mature. The first symptoms are small, brown, elliptical spots, usually less than 10 mm in diameter, on the surface of roots. The spots are often surrounded by a yellow halo. A section through the spots will show a small cavity a millimetre or so below the outer skin of the root. Cavity spot is particularly evident when the surface layer of the spot or lesion rots and disintegrates, revealing an open, sunken cavity up to 12 mm wide. Cavity spot develops rapidly on mature and over-mature carrots.

Infection of the taproot by *Pythium* species commonly causes the main root to fork or branch below ground level.

Source of infection and spread

Pythium species are soil-borne pathogens and persist as resting spores. *Pythium sulcatum* has a narrow host range

Fig 9.6 Forking in carrot roots.

restricted to carrot and related species, while *P. violae* has a wider host range that includes broadleaf weed species and some cereals.

Cavity spot is usually more severe in acid soils and in poorly drained, compacted soils.

Importance

Cavity spot is a common problem in carrot production in Australia, particularly in crops grown with limited rotation or harvested in late summer and autumn. The disease reduces the quality of carrots, making them unsuitable for both export and local markets.

Management

- Rotate carrots with unrelated crops such as brassicas, lettuce, capsicum and onions where *Pythium sulcatum* is present. When *P. violae* is present, rotate carrots with non-host crops that include cereals, onions, potatoes and beans. A five-year rotation is desirable.
- Grow carrot varieties that are tolerant to cavity spot.
- Maintain soil pH in the range pH 6.5–7. Avoid acid soils and those prone to waterlogging and compaction.
- Apply recommended chemicals at seeding or soon after to reduce *Pythium* damage.
- Harvest carrots as soon as they reach marketable size as over-mature carrots are more likely to develop cavity spot.

■ DAMPING-OFF

Refer to the chapter on Common diseases of vegetable crops.

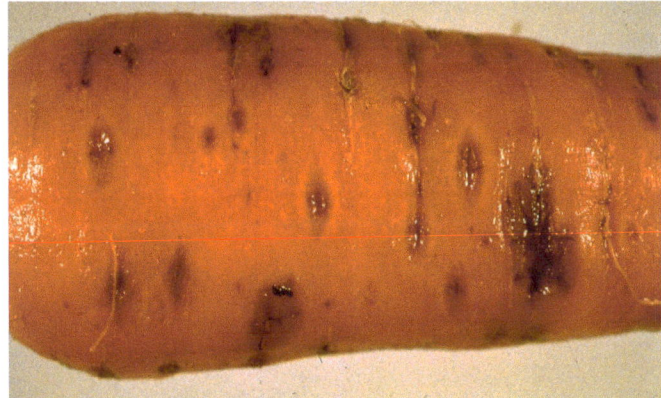
Fig 9.5 Cavity spot symptoms on a carrot root.

■ LEAF SPOT

Cause

The fungi *Alternaria dauci* and *Cercospora carotae*.

Symptoms

Alternaria leaf spot causes dark grey to brown, angular spots on leaves. When conditions favour infection, the lesions may coalesce and the whole leaf yellows, collapses and dies. The disease attacks older leaves first, but in severe outbreaks, younger leaves are also affected. Elongated lesions may also form on the petioles. *Cercospora* leaf spot is more severe on the young leaves developing initially as small necrotic flecks that enlarge to form circular, tan or grey spots. Spots may coalesce during humid weather to blight the entire leaf. Sunken, elongated spots may also occur on leafstalks.

Source of infection and spread

The fungi survive on undecomposed crop residues in the soil and are introduced on carrot seed. Spores produced on the leaves spread rapidly during wet, windy weather. Mild to warm temperatures are favourable for infection.

Importance

Leaf spot diseases can be serious in wet seasons.

Management

- Spray with the recommended fungicides. Base spray programs on regular applications of protectant fungicides (every seven to 10 days), limiting systemic and protectant tank-mixes to periods when conditions are favourable for disease development.

- Rotate carrots with other crops.

- Plant only high quality seed. Do not plant seed potentially infected with *Alternaria* or *Cercospora*.

■ POWDERY MILDEW

Cause

The fungus *Erysiphe heraclei*.

Fig 9.7 Foliar symptoms of Alternaria leaf spot.

Fig 9.8 Field symptoms of powdery mildew.

Fig 9.9 A close view of powdery mildew on carrot.

Symptoms

White powdery fungal growth first appears on the oldest leaves but quickly spreads to cover all leaf surfaces, as well as petioles, flower stalks and bracts. When infection is severe, the leaves become chlorotic and may die prematurely. Severe infection can cause the foliage to twist and become malformed.

Source of infection and spread

Erysiphe heraclei has a wide host range, affecting many plants within the Apiaceae family. Alternative hosts may be an important source of inoculum. In Australia, the fungus is also weakly pathogenic to parsnip and parsley. The spores are wind-borne and may be carried long distances in the air. High humidity and moderate temperatures favour infection and disease development. Spores develop between seven and 14 days after infection, when conditions are favourable for disease.

Importance

Powdery mildew occurs in most production areas in Australia.

Management

- Spray with the recommended fungicides.

- Destroy old crops promptly.

- Rotate carrots with crops that do not belong to the Apiaceae family.

■ SCLEROTINIA ROT

Refer to the chapter on Common diseases of vegetable crops.

■ NEMATODES

CARROT CYST NEMATODE – BIOSECURITY THREAT

Cause

The nematode *Heterodera carotae*.

Symptoms

Affected fields have areas of stunted, weak plants with yellowish-red foliage. Below ground, the taproots are smaller than normal and have a proliferation of lateral feeder roots, giving plants a beaded appearance. The nematode can be detected by examining roots for numerous brown cysts and white, pin-head sized females.

Life cycle

The carrot cyst nematode has a narrow host range that is restricted to cultivated and wild carrot species, and the weed species *Torilis arvensis*. Second stage juveniles emerge from eggs and move through the soil to parasitise the fine roots. The nematode spreads in contaminated soil on farm implements, with soil movement and in cysts on carrot roots.

Fig 9.10 Carrot cyst nematode infection, showing the proliferation of fine feeder roots.

Fig 9.11 Plants affected by carrot cyst nematode (left).

Importance

The nematode occurs in the United Kingdom, Europe, Cyprus, India and Michigan State in the USA. It is not present in Australia and is considered a major threat to the industry. Once established, the nematode is difficult to manage and causes significant crop losses.

What to do if you suspect carrot cyst nematode

This pathogen is a biosecurity risk to Australia. Any suspected affected plants should be reported to the nearest Department of Primary Industries or the Plant Health Australia hotline (1800 084 881).

■ ROOT-KNOT NEMATODE

Refer to the chapter on Common diseases of vegetable crops.

Fig 9.12 Root-knot nematode symptoms on carrot.

Fig 9.13 Root-knot infection causing stunting in carrot plants (foreground).

VIRUSES

■ *CARROT VIRUS Y*

Cause

Carrot virus Y (CaVY; Potyvirus)

Fig 9.14 Foliar symptoms of plants infected with *Carrot virus Y*.

Importance

The virus occurs in several Australian States and can cause serious losses on properties growing carrots continuously through the year, which facilitates its spread between consecutive crops.

Management

- Avoid planting sequential crops next to each other and separate crops spatially by planting other non-host crops between them.

- Destroy harvested crops and volunteer carrots promptly.

- An annual period free from carrot production will break the disease cycle.

■ MOTLEY DWARF DISEASE

Cause

Carrot mottle virus and *Carrot red leaf virus*.

Symptoms

Affected plants lose vigour and become pale and distorted. Symptoms are most serious in winter with tops dwarfed and flattened against the ground and prominently coloured yellow and purple. Yield is reduced.

Source of infection and spread

The virus complex is transmitted by aphids, particularly the carrot aphid (*Cavarilla aegopodii*).

Importance

Motley dwarf was once a serious disease of carrots in Australia, but rarely seen now, as most varieties are resistant.

■ FURTHER INFORMATION

Cunnington JH, Watson A, Liberato JR and Jones RH (2008). First record of powdery mildew on carrots in Australia. *Australasian Plant Disease Notes* **3**, 38–41.

Davis RM and Raid RN (Eds) (2002). *Compendium of Umbelliferous crop diseases.* APS Press, St Paul, Minnesota.

Fig 9.15 Root symptoms associated with *Carrot virus Y* infection (right).

Symptoms

Leaves on infected plants have a feathery appearance and develop a chlorotic mottle, reddening and general chlorosis. Taproots on young, infected plants are stubby, severely distorted and knobbly.

Source of infection and spread

The host range of *Carrot virus Y* is limited to carrot and several closely related species such as parsnip, coriander.

Aphids spread the virus from plant to plant. Several species can efficiently transmit the virus with only very short feeding times required to transmit the virus from plant to plant.

CaVY is not known to be seed-borne and can only survive in living host plants.

Celery (*Apium graveolens* var. *dulce*), a member of family Apiaceae, is grown for its thick, edible petioles or leaf stalks. The plant's origin is not clear as wild forms occur in marshy areas throughout temperate Europe, North Africa and western Asia. Celery is grown usually as a cool season crop, when temperatures are between 13 and 24°C. Commercial crops are generally established as transplants.

Fungal leaf blights and *Celery mosaic virus* are the main disease problems in Australian celery crops.

BACTERIA

■ BACTERIAL BLIGHT

Cause
The bacterium *Pseudomonas syringae* pv. *apii*.

Symptoms
Leaf spots are initially bright yellow, turning brown in the centre as the spots enlarge. Spots are usually circular but may be angular in shape and are surrounded by a yellow halo. Leaves with many spots may be blighted and die.

Source of infection and spread
Seed-borne infection is likely to be the main source of initial inoculum. The bacterium can survive epiphytically on leaves until conditions favour development and plant infection through the stomata or wounds. The bacterium spreads in water splash and wind blown rain. Diseased seedlings can introduce the pathogen into production fields. A dense crop canopy provides wet, humid conditions that are ideal for infection and disease development. The bacterium can survive in diseased crop residues.

Importance
Bacterial blight is more likely to occur in transplant nurseries on lush plants growing under warm, humid conditions.

Management
- Sow clean seed or treat with hot water if necessary.
- Maintain a high standard of hygiene in nurseries.

Fig 10.1 Field symptoms of bacterial blight.

Fig 10.2 Bacterial blight on celery leaves.

- Avoid leaving plants wet for prolonged periods.
- Destroy crop residues promptly after harvest.
- Rotate celery with other crops.

FUNGI

■ EARLY BLIGHT

Cause
The fungus *Cercospora apii*.

Symptoms
Small, yellow spots develop on leaves and enlarge rapidly to form irregularly rounded, orange-grey lesions. The texture of affected areas becomes dry and papery. In wet weather, a very fine, ashen-grey fungal mould develops on the spot. Symptoms appear first on the outer leaves.

Elongated spots may also develop on the leafstalk when conditions favour the disease.

Source of infection and spread
The fungus survives on undecomposed celery residues in the soil and may also be introduced in seed. Fungal spores are blown long distances by wind and spread in splashing water and on people and implements moving through a crop. Heavy, persistent dews and showery, humid weather favour the disease.

Importance
Early blight is a common and serious disease if it is not controlled.

Management
- Treat seed in hot water or with fungicides to eliminate seed-borne infection.
- Apply the recommended fungicides to crops before the diseases become serious.
- Use crop rotation to allow crop residues to decompose before replanting to celery.
- Use furrow or drip irrigation or time overhead irrigations to allow plants to dry during the day.
- Use wider row-spacing and raised beds to improve air circulation.

■ LEAF CURL

Cause
The fungi *Colletotrichum acutatum* and *Colletotrichum orbiculare*.

Symptoms
Plants of any age may become infected. Leaves develop a distorted, bent appearance, and curl downwards. These leaves later become very brittle, and cracking occurs along their lengths. Small, translucent spots may develop on the leaf blade, and light brown, elongated lesions appear on the petioles. Soft-rotting organisms often invade the infected growing point, effectively preventing further growth.

Source of infection and spread
The fungi may survive on undecomposed celery residues in the soil or on numerous alternative crop and weed hosts such as strawberry, capsicum, papaya, avocado, Noogoora burr (*Xanthium pungens*) and Bathurst burr (*Xanthium spinosum*). Spores spread rapidly during wet, windy weather.

Fig 10.3 Early blight caused by *Cercospora apii*.

Fig 10.4 Field symptoms of leaf curl disease.

Fig 10.5 Leaf curl causing death of the growing tip.

Importance
Leaf curl may be serious in susceptible varieties following warm, wet weather.

Management
- Use resistant varieties.
- Rogue infected plants to prevent spread.
- Apply the recommended fungicides in both the seedling propagation area and the field.

Fig 10.6 Leaf curl. Note distortion in the stalks.

Fig 10.7 Leaf curl showing vertical and horizontal cracks.

- Destroy old crops promptly.
- Avoid planting into soil containing infested crop residues.

■ SCLEROTINIA ROT
Refer to the chapter on Common diseases of vegetable crops.

■ SEPTORIA SPOT (LATE BLIGHT)

Cause
The fungus *Septoria apiicola*.

Symptoms
The disease is first seen on the lower, older leaves as small, yellowish spots that enlarge and turn brown. The spots become studded with small, black fruiting bodies (pycnidia). Affected leaves appear 'fired' as the spots coalesce, the leaf withers and dies. Greyish-brown, irregularly shaped spots may also develop on the leaf stalks.

Source of infection and spread
The fungus survives on undecomposed celery residues in the soil and may also be introduced on celery seed. The pathogen can survive for two to three years on seed when stored under dry, cool conditions. Spores produced in fruiting bodies spread during wet, windy weather or in overhead irrigation. Cool, moist weather resulting in long periods of leaf wetness favours infection and disease development.

Importance
Septoria spot is a common and often serious disease of celery.

Fig 10.8 Septoria spot (late blight) in celery.

Management

- Treat seed in hot water or with fungicides to eliminate seed-borne infection.

- Apply the recommended fungicides to crops before the diseases become serious.

Fig 10.9 A leaf affected by Septoria spot. Note the small, black, fungal fruiting bodies on the lesions.

Fig 10.10 Advanced symptoms of Septoria spot (late blight).

- Use crop rotation to allow crop residues to decompose before replanting to celery.

- Use furrow or drip irrigation or time overhead irrigations to allow plants to dry during the day.

- Use wider row-spacing and raised beds to improve air circulation.

VIRUSES

■ CELERY MOSAIC

Cause

Celery mosaic virus (CeMV; Potyvirus).

Cucumber mosaic virus (CMV; Cucumovirus)

Symptoms

Celery mosaic virus causes vein clearing and mottling on the younger leaves. On older foliage, leaflets become narrow,

Fig 10.11 *Celery mosaic virus.*

Fig 10.12 Celery mosaic symptoms.

twisted and cupped. Leaf size may reduce and plants become stunted with a flattened appearance. Dark green mottled areas can develop on leaf stalks.

Cucumber mosaic virus (CMV) also infects celery, causing similar symptoms to *Celery mosaic virus*, although the latter does not cause brown, sunken spots on leaf stalks sometimes associated with CMV infection. Laboratory tests may be necessary to determine the causes of symptoms in crops.

Source of infection and spread

Aphids spread both *Celery mosaic virus* and *Cucumber mosaic virus* from plant to plant. A large number of aphid species can transmit both viruses with only very brief feeding periods of approximately one minute being required for transmission. Neither virus is not carried in celery seed. CeMV can infect several weed and cultivated species in the Apiaceae family, although it seems largely restricted to celery in Australia. The main source of the virus is infected celery and serious outbreaks generally occur where celery is grown throughout the year with sequential plantings of crops in close proximity. This allows aphids to carry the virus over very short distances to young plantings.

CMV has a very wide host range among crop and weed species.

Importance

CeMV can be a major problem where celery crops are grown throughout the year with sequential plantings.

Management

- Produce seedling plants away from production areas and maintain a high standard of hygiene within and around the nursery.

- Avoid planting new crops next to older crops that may have high levels of virus infection and provide a source of virus for aphid transmission.

- Destroy harvested crops promptly and control volunteer celery plants and weeds in fallow areas and headlands.

- Celery-free periods of one to three months have been successful in controlling major outbreaks in several countries.

■ FURTHER INFORMATION

Davis RM and Raid RN (Eds) (2002). *Compendium of Umbelliferous crop diseases*. APS Press, St Paul, Minnesota.

Cucurbits belong to the family Cucurbitaceae and comprise frost-sensitive, tendril-bearing vines growing predominantly in subtropical and tropical regions. The most important cultivated species in Australia are cucumber (*Cucumis sativus*), rockmelon and honeydew (*Cucumis melo*), watermelon (*Citrullus lanatus*), pumpkin (*Cucurbita maxima*) and squash and zucchini (*Cucurbita pepo*).

Species of Cucumis, Momordica, Tricosanthes and Lagenaria are grown as non-leafy Asian vegetables and are susceptible to the same range of pathogens as the major cucurbit types.

BACTERIA

■ ANGULAR LEAF SPOT

Cause
The bacterium *Pseudomonas syringae* pv. *lachrymans*.

Symptoms
The early symptoms are the development of irregular or angular water-soaked leaf spots. They become angular in shape as they expand and the edges become delimited by the veins. As the leaf spots age they become dry and grey in

Fig 11.2 Angular leaf spot on a melon fruit.

colour. The dry tissue often tears and falls out, giving a tattered appearance.

Small, circular, water-soaked spots occur on fruit, becoming dry and almost white in colour.

Source of infection and spread
The bacterium is seed-borne and may also survive from season to season on crop residues. The bacterium spreads

Fig 11.1 Angular leaf spot on a rockmelon leaf.

Fig 11.3 Angular leaf spot on a cucumber leaf.

in wind-blown rain or in overhead irrigation, and warm, wet weather favours the disease.

Importance

The disease may be important under warm, humid conditions with cucumber and rockmelon more likely to be affected.

Management

- Rotate crops.

- Acidified hot water treatment (50°C for 20 minutes) of seed reduces seed-borne infection.

- Avoid high levels of nitrogen fertiliser as excessive nitrogen may increase susceptibility to disease.

- When conditions favour disease development, adopt a preventative spray program using copper and mancozeb tank-mixes.

- Remove and destroy affected crops promptly after harvest.

- Remove and destroy volunteer cucurbits, particularly in fields with a history of the disease.

■ BACTERIAL SOFT ROT

Cause

Bacteria of the genus *Erwinia* spp.

Symptoms

Soft, wet rots develop during transit and storage of postharvest fruit in a wide range of cucurbits (refer to the chapter on Common diseases of vegetable crops). The

Fig 11.4 Bacterial soft rot of zucchini fruit.

bacteria are also responsible for preharvest symptoms, primarily in zucchini crops. A brown, wet rot develops in the plant crowns and stems. Developing fruit are infected from the blossom end resulting in soft, slimy and rapid breakdown of fruit tissues. Severe infection in plant crowns results in a general plant wilt. A foul odour is associated with the rotten stem and fruit tissues.

Source of infection and spread

The bacterium is a ubiquitous organism that is common in decaying plant material and crop debris in the soil. Infection occurs through wounds in the plant tissues caused by mechanical damage or insect feeding. Hot, wet weather favours infection and the disease develops rapidly following periods of heavy rain or high humidity. Secondary spread within fields can occur via water splash, contaminated harvest equipment, farm workers, and insects.

Importance

The disease may be problematic in zucchini crops grown in tropical production areas, particularly following periods of heavy rain and warm weather.

Management

- When conditions favour disease development, adopt a preventative spray program using copper and mancozeb tank-mixes.

- Avoid harvesting fruit when the plants are wet, as secondary spread of the pathogen will occur most easily in water-splash between affected and non-infected plants.

- Disinfect harvesting tools regularly with a general agricultural sanitiser, particularly when moving between diseased and healthy crops.

- Remove and destroy affected crops promptly after harvest.

- Remove and destroy volunteer cucurbits, particularly in fields with a history of the disease.

■ BACTERIAL SPOT

Cause

The bacterium *Xanthomonas campestris* pv. *cucurbitae*.

Symptoms

Small, tan-yellow, angular spots, 1–2 mm wide, develop on leaves. Under wet conditions, spots appear greasy on the undersides. With time, the lesions become more rounded, and brown/translucent, with a wide yellow halo. The

Fig 11.5 Bacterial spot on a pumpkin leaf.

lesions enlarge, reaching diameters of up to 7 mm. If a number of spots occur together, that section of the leaf may brown and die. Infection frequently occurs at the leaf margin, causing tissue death.

Infection of young, immature fruit causes small water-soaked areas with pronounced, light brown ooze. As the fruit enlarge, the ooze dries to form a raised, yellow crust over the spot.

On mature fruit, spots may be up to 10 mm wide with a dark green, greasy margin. Young fruit are more susceptible, being relatively soft and without the resistant wax layer which develops on the surface of older fruit. The flesh beneath the spots may be dark and water-soaked, extending into the seed cavity and resulting in seed contamination.

Source of infection and spread

The bacterium is commonly introduced in seed. Once established, it spreads rapidly during wet, windy weather and in overhead irrigation. The bacterium may survive on undecomposed crop residues.

Importance

Bacterial spot is a sporadic disease but it can cause severe loss in pumpkins. Affected fruit are unmarketable, and secondary rots may cause complete loss. The disease occurs less frequently in rockmelon, watermelon and zucchini.

Management

- Plough in diseased crops immediately after harvest.

- When conditions favour disease development, adopt a preventative spray program using copper and mancozeb tank-mixes.

- Remove and destroy volunteer cucurbits, particularly in fields with a history of the disease.

■ BACTERIAL FRUIT BLOTCH

Cause

The bacterium *Acidovorax avenae* subsp. *citrulli*.

Symptoms

In Australia, this disease is predominantly a problem in rockmelon, honeydew and watermelon, while also recorded on cucumber and pumpkin. Seedling plants of watermelon, rockmelon and honeydew develop dark, angular, water-soaked areas on the cotyledons or seed-leaves. On true leaves, the leaf veins delimit the lesions. After several days, the affected areas may collapse and become pale brown and dry. Under warm, humid conditions, the disease may progress causing seedling death.

On more mature plants, leaf infection may often be minor. Leaf lesions are initially dark brown, angular and water-soaked and tend to elongate along leaf veins. If warm, humid or wet conditions persist, the lesions produce abundant

Fig 11.6 Bacterial spot on mature pumpkin fruit.

Fig 11.7 Bacterial fruit blotch on rockmelon seedlings showing characteristic, angular, water-soaked lesions.

Fig 11.8 Blotch symptoms on cotyledons and young watermelon leaves. Note the dark, angular lesions.

bacterial ooze that appears as a white, chalky deposit on the leaves under the lesions. With age, the lesions dry out, becoming light brown and papery. The centres of old lesions often tear and drop out, giving the leaf a ragged appearance.

Infection of fruit tends to occur early in fruit development before the fruit have developed a tough outer skin. Sometimes, if drier conditions prevail during fruit set, the

Fig 11.9 Lesions on the underside of a leaf caused by bacterial fruit blotch. Note the water-soaking along the leaf veins.

Fig 11.10 An advanced infection of bacterial blotch on a rockmelon leaf.

disease may exist in the crop as a foliar infection only and no fruit symptoms may develop. If infection occurs soon after pollination, fruitlets may abort. On fruit, symptoms vary depending on the crop.

Watermelon fruit: Distinctive symptoms develop on watermelon fruit. A dark olive-green stain or blotch develops on the upper surface of fruit. The symptoms first appear as small irregularly shaped spots that rapidly expand into large blotches, covering large areas of the fruit. Blotches are at first water-soaked and grey or dark green in colour. As the disease progresses, the older central area of the blotch can become necrotic and brown to red-brown in colour. The skin surface cracks and amber-coloured exudate may ooze from the centre of the blotch. Fruit

Fig 11.11 Fruit blotch on rockmelon fruit. Note there are minimal external symptoms and the break in the fruit netting.

blotches are usually superficial on the rind. At the initial site of infection, small tan or brown corky cavities can extend through the rind into the flesh. This allows the bacterium to contaminate the seed. These penetration sites are often minor and may be difficult to detect. Secondary decay organisms invade the fruit blotch areas and are responsible for the decay and collapse of the fruit.

Rockmelon fruit: Detection of fruit infection in rockmelon is often extremely difficult, with fruit often showing very limited external symptoms, even when internal symptoms are severe. Small cracks, pits or breaks in fruit netting and small water-soaked spots may be the only signs. Internally, tan-dark brown discoloured areas and corky cavities develop under the rind, in the flesh and penetrate into the seed cavity.

Honeydew fruit: As with rockmelon, detection of fruit infection may be extremely difficult. Small, translucent, water-soaked patches on the rind may be the only sign. Internal symptoms are similar to those seen in rockmelon.

Source of infection and spread

There are two strains of this pathogen in Australia. The 'watermelon strain' causes disease on both honeydew and watermelon but is less pathogenic to rockmelon, while the 'rockmelon strain' is strongly pathogenic to all three hosts. The bacterium is seed-borne and serious outbreaks of the disease are often associated with seed contamination. Disease development in seedling nurseries and melon crops is favoured by warm, humid conditions. The bacterium can be introduced into production fields on infected transplants and secondary spread occurs through water splash and on machinery, clothing, and footwear.

The bacterium can also survive on volunteer melons, weed species within the cucurbit family, in particular prickly paddy melon (*Cucumis myriocarpus*) and crop residues.

Fig 11.13 Severe skin and internal symptoms of blotch in melon fruit Note the brown, corky lesions.

Young fruit are more susceptible to infection and the bacterium enters the immature fruit through stomata before the protective waxy layer (watermelon) or corky netting (rockmelon) has formed.

Importance

Bacterial fruit blotch is a major disease of cucurbits worldwide. It can cause severe losses in seedling nurseries and melon production areas when warm wet weather favours the spread of seed-borne infection.

Management

- Use seed tested and free of seed-borne contamination.

- Do not plant seed known to be contaminated. If seed is untested for the disease, treat it in hot water (25 minutes at 55°C) to reduce or eliminate internal and external contamination. After treatment, cool seed quickly with running water, dry and sow it within two days. Heat-treat seed with caution as some varieties or seed lots are affected adversely by hot water treatment.

Fig 11.12 Fruit blotch infection in watermelon.

Fig 11.14 External blotch symptoms on watermelon.

BACTERIAL WILT – BIOSECURITY THREAT

Cause
The bacterium *Erwinia tracheiphila*.

Symptoms
Initially the disease appears as a green wilt of individual leaves or plant runners or a wilt of the entire plant. As the disease develops the leaves may become chlorotic and then necrotic and the plant may collapse and die. Cucumber and melons are affected most severely, pumpkins less so and watermelons are apparently not susceptible. A transverse cut of a wilted runner near the crown of the plant will reveal strands of a gummy exudate that forms sticky threads between the cut stem surfaces when the two parts of the stem are drawn apart.

Fig 11.16 Leaf symptoms of bacterial wilt.

Source of infection and spread
Mechanical transmission via cucumber beetles is the principle method of spread within and between fields. *Erwinia tracheiphila* is host-specific to members of the Cucurbitaceae and beetles that contact or feed on infected plants become contaminated with the bacterium and transmit it to healthy plants during feeding. The bacterial cells multiply at feeding wound sites and infect the plant systemically via the plant xylem vessels. Wilting occurs when the vascular cells become plugged with resins and masses of bacteria. The source of initial infection is not well understood but is likely to be infected cucurbit weeds or volunteers. The disease is not seed-borne and it does not persist in the soil in the absence of a host.

Importance
This disease is most serious in the United States, but it is also found in several Asian and African countries and it has been reported in Europe. The presence of a suitable insect vector is important for the spread of this pathogen. Neither the insect vectors of the pathogen nor the disease has been reported in Australian cucurbits.

What to do if you suspect bacterial wilt
This pathogen is a biosecurity risk to Australia. Any suspected affected plants should be reported to the nearest Department of Primary Industries or the Plant Health Australia hotline (1800 084 881).

Fig 11.15 Green wilt in squash vines infected with bacterial wilt.

Fig 11.17 Bacterial wilt, showing chlorotic and necrotic lesions at the leaf margin. The cucumber beetle transmits the *Erwinia* bacterium.

- In transplant nurseries, practice good hygiene, reduce humidity levels and avoid overcrowding trays. Avoid using overhead sprinklers late in the day and dispose of diseased plants promptly.

- Remove volunteer and weed cucurbits (particularly prickly paddy melon) from around nurseries and production fields.

- Use a three-year rotation for melon production areas.

- When environmental conditions favour infection, initiate a preventative program of copper sprays well before fruit set.

FUNGI

■ ALTERNARIA FRUIT ROT

Cause
The fungus *Alternaria alternata*.

Symptoms
Circular to oval lesions develop on fruit and may become sunken. Affected skin first appears bleached or brown but is covered rapidly by a dark mould with abundant olive-green to dark grey spores.

Source of infection and spread
The fungi survive on crop residues in the soil and may also be introduced on seed. The fungus colonises dead tissues of petioles and stems and may also be a surface contaminant. Spores spread by air movement, with warm, moist weather favouring the disease. Fruit damaged by sunscald or weakened by prolonged storage, or chilling are prone to *Alternaria* infection.

Importance
Alternaria fruit rot is a relatively common cause of loss in rockmelon. The disease often manifests as a post-harvest problem. Other cucurbits such as squash, cucumber, honeydew and watermelon may also be affected.

Management
Refer to the management of sour rot below.

■ ALTERNARIA LEAF BLIGHT

Cause
The fungus *Alternaria cucumerina*.

Symptoms
Small, tan-coloured spots develop on leaves, enlarging to roughly circular, brown areas that may have concentric ring markings. Spots may coalesce to cover almost the entire leaf.

Source of infection and spread
The fungus survives on crop residues in the soil and may also be introduced on the seed. Spores spread in the wind, and warm, moist weather favours the disease.

Importance
The disease affects the majority of cucurbits, but is more common on pumpkin and cucumber. *Alternaria* leaf blight may be a significant problem during warm wet weather.

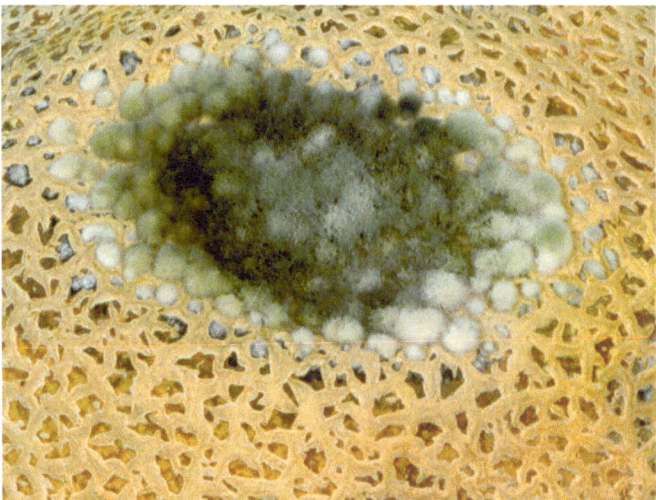

Fig 11.18 Alternaria fruit rot of rockmelon.

Fig 11.19 Alternaria leaf blight.

Fig 11.20 Variation in Alternaria leaf blight symptoms.

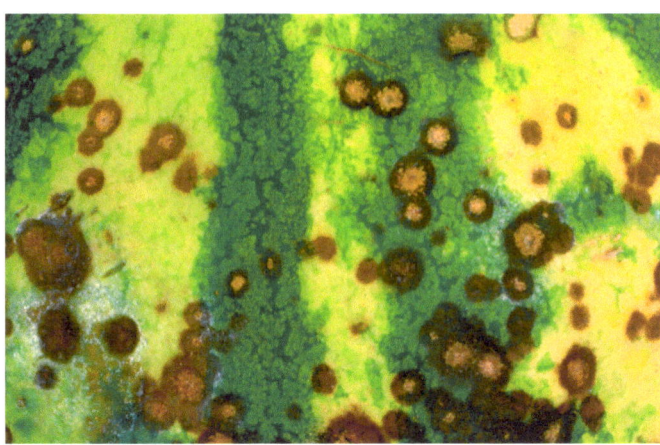

Fig 11.22 Anthracnose on watermelon fruit.

Management

- During periods of warm, humid or wet weather, adopt a preventative fungicide spray program with the recommended protectant fungicides.

■ ANTHRACNOSE

Cause

The fungus *Colletotrichum orbiculare*.

Symptoms

Small, brown, circular areas surrounded by a yellowish halo develop on leaves. These enlarge to form circular to elongate, dark brown to black spots, often centred on the veins. Pale brown, slightly sunken, elongated areas develop on runners.

On fruit, circular, pale brown, depressed spots with a raised margin develop, usually concentrated on the lower half of the fruit. Spots are up to 30 mm wide, but they often coalesce to cover large areas of the fruit. Pink to orange spore masses develop on the spots in moist weather. Although the infection does not reach the flesh, secondary rots may develop, causing extensive decay.

Source of infection and spread

The fungus survives on crop residues, volunteer cucurbits and may be carried on the seed. The orange to pink spores produced on affected parts spread in splashing water and on contaminated implements, clothing and footwear. Warm, humid weather is necessary for infection and disease development, with spores requiring 100% relative humidity for 24 hours to germinate and infect host tissue.

Importance

Seven races of this pathogen have been identified worldwide, however the prevalence of specific races in Australia is not known. The disease can cause serious losses in melon and

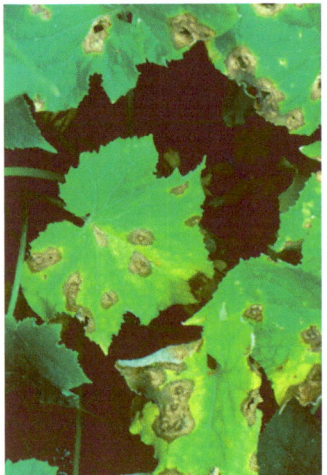

Fig 11.21 Anthracnose on cucumber leaves.

Fig 11.23 Anthracnose on honeydew melon.

cucumber crops, if frequent showery weather occurs from fruit set to maturity. Leaf symptoms also occur on rockmelon and cucumber. Anthracnose was once a very serious disease but generally, it is now well controlled through healthy seed.

Management
- During periods of warm, humid/wet weather, adopt a preventative, fungicide spray program using the recommended protectant fungicides.

■ BLUE MOULD ROT

Cause
The fungi *Penicillium* spp.

Symptoms
Melons and cucumbers are most likely to be affected by this disease. Circular, water-soaked lesions develop on fruit. The diseased areas often rupture and are covered by blue-green spores. Affected fruit has a typical musty odour.

Source of infection and spread
Fungal spores are abundant in soil and in the atmosphere. Fruit is contaminated during harvesting or packing, with infection usually occurring through wounds or directly through skin that has been weakened by sunburn, chilling injury or prolonged storage.

Importance
Blue mould rot occurs throughout the world. Cucumber and melon fruit are most likely affected.

Management
- Maintain strict hygiene in packing sheds, removing diseased fruit daily and disinfect the shed and equipment.
- Handle fruit carefully to minimise injuries.
- Harvested fruit should not be stored for long periods.

■ BROWN OR SURFACE ETCH

Cause
The fungi *Fusarium* spp. and *Didymella bryoniae*.

Symptoms
The disease commonly affects the butternut pumpkin and occasionally other pumpkin varieties. Distinctive bronze areas with concentric bands develop on the fruit, mostly on the lower surface. These areas may be quite large and often crack in the centre. The bronze colour changes to a light brown-grey with age. Only the skin of the fruit is affected, but if secondary rots enter through the cracks, complete breakdown may occur. Symptom development ceases once fruit is mature.

Source of infection and spread
Fungi causing brown etch are soil inhabitants and invade the fruit where ground contact has occurred. The disease is favoured by warm, wet weather.

Importance
The disease can be serious in butternut pumpkin.

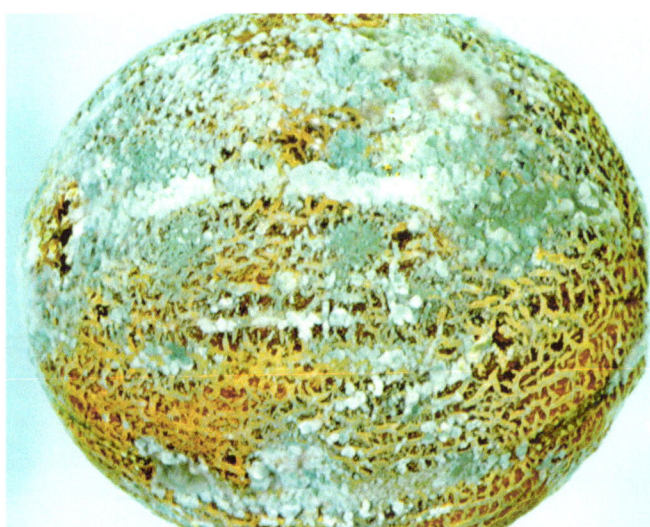

Fig 11.24 Blue mould rot on fruit.

Fig 11.25 Brown etch on butternut pumpkin.

Management

- Use less susceptible varieties in badly infested areas and during the summer.

- Rotate cucurbits with other crops.

- Increase calcium levels in plants and fruit as this will increase skin elasticity and reduce minute skin cracks, which provide entry points for the pathogens.

- Wash fruit after harvest, using a recommended sanitiser to reduce inoculum levels.

- Avoid overhead irrigation systems.

■ CHARCOAL ROT

Cause
The fungus *Macrophomina phaseolina*.

Symptoms
Older leaves near the crown wilt, yellow and eventually die. The crown and lower stem develop a dark green water-soaked lesion that can eventually cover the entire crown and extends above the ground. Brown exudates occur on the surface of lesions. As the disease progresses, the stem and crown lesions will develop a dry, tan appearance the stem tissue cracks. Minute, black microsclerotia may be visible embedded in the diseased tissue. Vines appear stunted or wilted and eventually collapse and die. Fruit in contact with soil can also develop a firm decay.

Source of infection and spread
The fungus is soil-borne and survives as microsclerotia in soil and crop residues. The fungus has a wide host range among crop and weed species. Infection occurs through the roots of plants with the fungus progressing to the crown. Symptoms usually appear as the fruit matures.

Importance
Although all cucurbits may be affected, charcoal rot is more likely to be a problem in melon crops grown under warm to hot conditions.

Management

- Rotate cucurbits with non-host crops to reduce fungal inoculun levels.

- Some melon varieties have some tolerance to the disease.

■ DOWNY MILDEW

Cause
The oomycete *Pseudoperonospora cubensis*.

Symptoms
Infected leaves develop small, pale yellow areas that enlarge and dry out to form brown, angular spots, often limited by the veins. The angular shape is most pronounced on cucumber, whereas spots are more rounded on rockmelon and zucchini. In humid weather, a purplish grey, downy growth may be visible on the undersides of spots. Affected leaves curl, shrivel and die.

Source of infection and spread
The organism is an obligate parasite and survives on volunteer cucurbits or weed species within the cucurbit family. Spores (sporangia) develop under high humidity and

Fig 11.26 Charcoal rot infection in rockmelon vines.

Fig 11.27 Advanced symptoms of downy mildew on a zucchini leaf.

Fig 11.28 Downy mildew on a cucumber leaf.

Fig 11.30 Field infection of downy mildew. A healthy plant is shown in the left foreground.

disperse in wind or water splash. Free moisture is required for spore germination and disease development is rapid under warm, humid conditions occurring during wet, showery weather or periods of fog or heavy dew. Temperatures between 18 and 23°C are optimal for infection whereas temperatures above 28°C are unfavourable for infection.

Importance
The majority of cucurbits are susceptible, but the disease can be particularly severe in susceptible varieties of cucumber, rockmelon and zucchini. Downy mildew is difficult to control in wet weather.

Management
- Apply the recommended protectant and eradicant fungicides. Base the spray program on regular protectant fungicide applications, using eradicant products only when environmental conditions are highly favourable for infection, or at the stage of early symptom development.

- Use resistant varieties.

- In protected cropping, reduce humidity levels and the presence of free water on the leaves.

■ FUSARIUM FOOT ROT

Cause
The fungus *Fusarium solani*.

Symptoms
In seedling plants, leaves become pale and wilt. Bases of affected plants are orange-brown to reddish-brown in

Fig 11.29 Downy mildew on a rockmelon leaf.

Fig 11.31 Fusarium foot rot, showing a healthy plant on the left.

colour. Roots initially appear healthy, but in advanced cases, they turn brown. Symptoms in older plants are similar. A characteristic of the disease is that soft tissues at the base of the plant disintegrate, leaving only the stringy water-conducting fibres that are usually reddish-brown in colour.

Source of infection and spread
Spores of the fungus can be carried on seed and the disease may occur in seedling nurseries. Once introduced to soil through the planting of infected seed or seedlings, the fungus can remain for one to two years. It spreads from plant to plant as spores carried by wind or running water.

Importance
Foot rot occurs infrequently, with outbreaks usually related to a particular seed source or inadequate rotation of crops. Pumpkin and rockmelon are usually the most severely affected.

Management
• Allow at least two years before replanting cucurbits in areas where the disease has occurred.

■ FUSARIUM FRUIT ROT

Cause
The fungi *Fusarium* spp.

Symptoms
Symptoms usually occur at the stem end and begin as small, scattered spots on the skin. Affected tissue tends to be spongy or corky and, as the disease progresses, becomes covered with a white or pinkish mould. A distinct margin usually occurs between healthy and affected tissue.

Source of infection and spread
The fungi are common soil inhabitants and often infect fruit on the underside. However, spores may be splashed on to any part of the fruit during rain or irrigation. Some form of wounding is necessary for infection. Hot, wet weather, particularly during harvest, favours rapid breakdown of fruit.

Importance
Fusarium fruit rot is a common cause of post-harvest loss, particularly in rockmelons.

Management
Refer to the chapter on Common diseases of vegetable crops.

■ FUSARIUM WILT

Cause
Fusarium oxysporum f.sp. *niveum* (watermelon), *Fusarium oxysporum* f.sp. *melonis* (rockmelon) and *Fusarium oxysporum* f.sp. *cucumerinum* (cucumber).

Symptoms
In seedlings, the cotyledons or seed leaves lose their healthy lustre and wilt, followed by the complete collapse of the plant. Older plants first show leaves wilting and yellowing near the crown. Later, individual runners are affected, and then the whole plant. Some recovery may occur at night, but the plant finally dies. A red-brown discolouration of the water-conducting tissues can be seen in the split stem and taproot. Fruit from affected vines are small, with poor flavour and colour.

Fusarium oxysporum cucumerinum causes a stem rot in glasshouse-grown cucumbers. Infected stems become

Fig 11.32 Fusarium fruit rot in rockmelon.

Fig 11.33 Early symptoms of leaf yellowing and wilting caused by the Fusarium wilt fungus, near the crown of a rockmelon plant.

Fig 11.34 Fusarium wilt on glasshouse-grown cucumbers.

covered with powdery orange spore masses. Maggots of sciarid flies (fungus gnats) can become numerous in the crown and larger roots as the disease develops.

Source of infection and spread

The fungus is a soil inhabitant that enters the plant through the root and grows into the water-conducting tissues. It also may be carried on seed, in soil adhering to implements and in contaminated water. The disease is favoured by warm to hot weather, and the fungus persists for long periods in the soil as a saprophyte or a resistant chlamydospore.

Importance

In Australia, the disease is most serious in watermelons, but considerable losses can also occur in rockmelons and cucumbers. Three races of *Fusarium oxysporum* f.sp. *niveum* have been reported worldwide on watermelon, four races of *F. oxysporum* f.sp. *melonis* have been recorded on rockmelon,

and three races of *F. oxysporum* f.sp. *cucumerinum* have been recorded on cucumber. The prevalence of individual races of the three pathogens in Australia is not known.

Management

- Plant resistant varieties.
- Practice at least three-year crop rotations so that pathogen populations can decline.
- Practice good farm and crop hygiene.

■ GROUND ROTS

Cause
Several fungi including *Rhizoctonia solani*, *Sclerotium rolfsii* and *Fusarium* spp.

Symptoms
Diseased areas develop where fruit contact the soil. Symptoms vary with the fungus involved.

Rhizoctonia and *Fusarium.* Small, circular, water-soaked spots develop, enlarging to form light brown to tan sunken areas with surface cracking.

Sclerotium. Large decayed areas with a prominent thick, white fungal growth covering the affected areas.

Source of infection and spread
The fungi are common soil inhabitants. Warm, wet weather favours their development.

Importance
Ground rots can cause pre- and post-harvest losses of fruit, particularly in rockmelon and cucumber.

Fig 11.35 Stem symptoms of Fusarium wilt on glasshouse-grown cucumbers.

Fig 11.36 Rhizoctonia ground rot.

Management

- Prepare land early so that crop residues thoroughly decompose before planting.

- Use plastic mulch to reduce fruit contact with soil.

- Do not market affected fruit.

■ GUMMY STEM BLIGHT

Cause

The fungus *Didymella bryoniae*.

Symptoms

The disease can affect all above-ground parts of the plant at any growth stage from seedlings to mature vines with fruit. Infection and symptoms can occur on all plant parts except roots. Symptoms on seedlings occur as light to dark brown spots on the cotyledons. Infection of the cotyledons or hypocotyl causes a water-soaked, brown discolouration of the tissues, followed by tissue desiccation and collapse. Seedlings die rapidly after infection of either the hypocotyl or cotyledons. On older plants, leaf symptoms appear as small circular, tan spots up to 5 mm in diameter, sometimes surrounded by a yellow halo. Under favourable conditions, the leaf lesions may enlarge rapidly and become irregular in shape. When the lesions coalesce, the entire leaf may become blighted. The spots dry, become cracked and may tear, giving leaves a tattered appearance. Infection often begins at the leaf margins.

Stem infections consist of brown oblong water-soaked lesions. Main stem lesions enlarge and slowly girdle the

Fig 11.38 Symptoms of gummy stem blight on leaves.

main stem resulting in a brown canker that produces a characteristic red or brown gummy fluid. Tiny black pimple-like fruiting bodies of the fungus (pycnidia) develop within the infected tissue. Cankers can girdle the entire stem and result in foliage wilting and affected areas dying. Vine wilting is usually a late symptom of the disease.

The most important form of the disease is crown rot, which may kill plants. At first, pale brown, then bleached areas develop, and a reddish gum oozes from cracks. Affected areas are studded with pycnidia of the fungus.

Fruit develop water-soaked, small oval to circular spots that are a greasy green colour and turn dark brown as the spots enlarge. Gummy exudate and black fruiting bodies may develop on the spots. Affected fruit may eventually become black in colour.

Fig 11.37 Necrosis and gummy exudate on rockmelon vines with gummy stem blight disease.

Fig 11.39 Advanced lesion of gummy stem blight on a vine showing exudate.

Fig 11.40 Gummy stem blight infection of seedling leaves.

Source of infection and spread
The fungus is seed-borne and can survive in soil, weeds and on crop residues. The fungal fruiting bodies contain large numbers of spores that spread in wind and splashing water. Warm, wet weather favours the disease.

Importance
Gummy stem blight is a major disease of cucurbits, particularly in tropical and subtropical areas. The disease can cause serious losses in watermelon, rockmelon, honeydew, squash, pumpkin and cucumber.

Management
- Rotate cucurbits with other crops on a two-year cycle.
- Apply the recommended fungicides, particularly if wet weather occurs.

- Destroy all organic debris from previous cucurbit crops by deep ploughing to reduce sources of inoculum carrying over to new plantings.

■ PINK MOULD ROT

Cause
The fungus *Trichothecium roseum*.

Symptoms
Rockmelon is the most common host for pink mould rot. Lesions usually form at the blossom end of the fruit and then extend over much of the surface. Affected skin becomes tough and shrivelled, and a viscous liquid may ooze from the lesions. A furry, pinkish, fungal growth often covers the affected area. Diseased flesh is spongy, slightly brown and has a very bitter taste.

Source of infection and spread
The fungus is a common soil inhabitant, and spores spread in moving air, irrigation water and on insects. The fungus invades melon fruit through soil contact, injuries, or at the cut stem during harvest. Warm weather favours the disease.

Importance
Pink mould rot is a minor disease but it can cause post-harvest losses in rockmelon.

Management
- Avoid injuries during harvest.
- Apply post-harvest treatments, if required.
- Cool and refrigerate fruit quickly after harvest.

Fig 11.41 Gummy stem blight on squash fruit.

Fig 11.42 Pink mould rot affecting rockmelon fruit.

■ POWDERY MILDEW

Cause

The fungus *Podosphaera xanthii*.

Symptoms

Small, circular, white, powdery patches occur on leaves, runners and leafstalks. These are generally visible first on the undersides of leaves, but eventually cover both surfaces. Yellow spots may form on upper surfaces opposite powdery mildew colonies on the lower surface. Affected leaves gradually turn yellow and become brown and papery.

Powdery mildew is most likely to develop in response to plant stress, for example stress associated with fruiting or the presence of other diseases.

Source of infection and spread

The fungus is an obligate parasite and cannot survive in the absence of a living host plant, except in special structures called cleistothecia, which protect the spores. Initial infecting spores may come from old cucurbit debris left in the field or be blown in on air currents from infected nearby fields.

Although powdery mildew occurs on a broad range of crops, each species is relatively host specific.

Podosphaera xanthii survives on volunteer and old cucurbit crops and possibly weed hosts, including weedy cucurbit species.

Fig 11.44 Field infection of powdery mildew on zucchini showing tolerant and susceptible plants.

The white mildew consists of large numbers of spores that spread by wind. Unlike downy mildew, this disease flourishes in comparatively dry weather, and spore germination is inhibited by free water on the leaf surface. Light dews provide sufficient moisture for infection. Initial infection is favoured by dense plant growth and low light intensity. The fungus occurs as several races, some of which can infect all cucurbits whereas others are restricted to one species. Strains resistant to eradicant fungicides also occur within the pathogen population.

Importance

Powdery mildew is a common and serious disease. Cucurbit fruits are not attacked directly by the powdery

Fig 11.43 A severe infection of powdery mildew on leaves.

Fig 11.45 Close view of powdery mildew fungal mycelium on the underside of a leaf.

mildew fungi, but early leaf loss can result in sunburnt fruit or uneven or premature ripening.

Management

- Spray with the recommended fungicides. Resistance to the highly active systemic fungicides may develop, but this is less likely to occur if they are used strategically in a program with protectant fungicides. Use recommended fungicide-resistance management programs to reduce the risk.

- Remove plant debris at the end of the season to reduce carryover of the fungus.

- Grow resistant varieties.

PHYTOPHTHORA BLIGHT – BIOSECURITY THREAT

Cause

The oomycete *Phytophthora capsici*.

Symptoms

The pathogen causes a seedling damping-off, and a crown and root rot as well as leaf spots, stem lesions and fruit rot. Initial symptoms include a rapid green wilt of the entire plant. Plants will often die within a few days of infection. The roots and stem tissues at the soil line turn brown and become soft and water-soaked and rapid stem collapse follows. Infected plants easily pull from the soil following extensive rot of the lateral roots. Fruit rot typically starts on the side of the fruit that is in contact with the soil. Large, dark brown water-soaked lesions can develop on leaves. Fruit symptoms begin as small water-soaked spots that may have visible fungal growth on their surfaces. Lesions become soft and easily puncture when handled and severe infection can cause fruit to collapse. Fruit symptoms can also develop rapidly post-harvest.

Fig 11.47 Phytophthora blight on zucchini. Note the water-soaked rot of the petiole bases.

Fig 11.46 Phytophthora blight on summer squash. Note the light-coloured mycelium and circular lesion on the fruit.

Fig 11.48 Phytophthora blight on cucurbit leaves. Note the large, brown, water-soaked leaf lesions.

Fig 11.49 Phytophthora blight of watermelon fruit, producing large water-soaked fruit lesions.

Source of infection and spread

The pathogen persists in soil as hard-walled resilient oospores and produces sporangia that give rise to motile zoospores that swim through soil moisture to invade susceptible plant roots. Spread of the spores in splashing water and surface water allows the pathogen to dissemination rapidly. The sporangia also disperse through the air. The disease is associated with warm temperatures (25–28°C), heavy rainfall, over-irrigation and poorly drained soils. The pathogen has a wide host-range across vegetable crops.

Importance

Phytophthora blight is a highly destructive disease of cucurbits. It is a major limitation to cucurbit production in parts of the United States.

What to do if you suspect Phytopthora blight

This pathogen is a biosecurity risk to Australia. Any suspected affected plants should be reported to the nearest Department of Primary Industries or the Plant Health Australia hotline (1800 084 881).

■ RHIZOPUS SOFT ROT

Cause

The fungus *Rhizopus stolonifer*.

Symptoms

Soft, water-soaked areas develop on fruit. The skin often cracks and coarse, white strands of fungus develop on the surface. The fungus later turns black, covering diseased areas with a mass of black spore heads (sporangia). Affected flesh is soft and wet. Rhizopus soft rot, in contrast to bacterial and *Geotrichum* soft rots, advances through the fruit on an even front.

Source of infection and spread

Rhizopus commonly occurs on plant debris in the soil. Spores are air-borne. Infection often occurs during wet weather when dying blossoms are invaded, allowing the fungus to enter the stem end of the fruit. Infection can also occur through wounds.

Importance

Rhizopus soft rot can be a serious post-harvest disease in rockmelons.

Management

- Avoid harvesting during wet weather.
- Handle fruit carefully to minimise injury.
- Cool fruit rapidly and as soon as possible after harvest.
- Remove diseased fruit from the vicinity of the packing shed.
- Apply the recommended post-harvest fungicide treatments.

■ SCAB

Cause

The fungus *Cladosporium cucumerinum*.

Fig 11.50 Rhizopus soft rot on rockmelon fruit.

Fig 11.51 Scab on cucumber fruit.

Symptoms

The fungus may attack any above-ground part of the plant. Circular to angular, water-soaked pale green spots develop on leaves. The spots enlarge and turn grey to brown in colour, often with a yellow margin. The inner tissue of the spot gradually breaks and falls giving a 'shot-hole' effect. If many spots occur on young leaves, they may become twisted and deformed. Under humid conditions, the spots on the leaves, stems and petioles may be covered with a dark green, powdery fungal growth.

Fruit infections begin as minute greasy sunken specks on the surface. These enlarge, remain sunken and become circular to oval in shape. A sticky exudate often oozes from the fruit lesions and secondary organisms invade causing fruit decay.

Source of infection and spread

The fungus survives for several years on crop residues and organic matter in the soil. It may also be carried on seed. Fungal spores disperse in moist air, and infection and disease development are favoured by cool, moist conditions including showery weather, heavy dews and fog.

Importance

The disease can affect most cucurbits including cucumber, squash, pumpkin and melons.

Management

• Use available, resistant varieties where the disease is likely to be a problem. Several cucumber varieties with scab resistance are available.

• Use seed free of the fungus.

• Rotate cucurbits with other crops with two or more years between cucurbits.

• Apply protectant fungicides, particularly if cool, wet weather is expected.

• Avoid planting in low-lying, shaded areas prone to heavy fogs and dews.

■ SEPTORIA SPOT

Cause

The fungus *Septoria cucurbitacearum*.

Symptoms

Small, brown leaf spots develop, with the centres eventually drying out and becoming studded with small, black fruiting bodies (pycnidia) of the fungus.

Lesions on fruit are small, circular, light brown, raised scabs, with star-shaped cracks occurring in the centre of mature lesions.

Source of infection and spread

The fungus survives on undecomposed crop residues. Large numbers of spores develop in the fungal fruiting bodies and spread in rain. Cool, wet weather favours the disease.

Importance

Septoria spot is mainly a disease of pumpkin, although found occasionally on marrow and squash. The disease is uncommon in Australia.

Management

• Fungicides used for other diseases should also control septoria spot.

Fig 11.52 Septoria spot on pumpkin. Light brown, raised spots on pumpkin fruit are caused by the fungus *Septoria cucurbitacearum*.

■ SOUR ROT

Cause
The fungus *Geotrichum candidum*.

Symptoms
Rockmelons are particularly susceptible to sour rot. The disease occurs after harvest and begins at the stem end. The flesh rots in preference to the skin and the fruit becomes a hollow shell containing an unpleasant-smelling liquid. A white, cheesy, fungal growth is often visible on decayed tissue. The disease is often confused with and may occur with bacterial soft rot.

Source of infection and spread
The fungus frequently originates in the soil and enters the fruit through small cracks in the stem end under wet conditions. After harvest, it spreads from fruit to fruit in contaminated wash water. Vinegar flies may also spread the fungus.

Importance
Sour rot is a major cause of post-harvest losses in rockmelon.

Management
- Avoid harvesting in wet weather.
- Handle fruit carefully to minimise injury.
- Cool fruit rapidly after harvest.
- Apply the recommended post-harvest fungicide treatments.

■ SUDDEN WILT OR VINE DECLINE

Also known as *Monosporascus* root rot.

Cause
An interaction between several pathogens (Australia: *Pythium* and *Fusarium* spp; elsewhere: *Monosporascus cannonballus*) and soil-plant-water relationships.

Symptoms
The disease is known by other common names that reflect the observed symptoms on infected plants. Vine decline is the most common symptom on rockmelon and honeydew melon, which are most susceptible to the disease. The vines appear to be developing normally, then suddenly turn yellow, wilt and die prematurely from fruit set. The rapid wilt of plants may result in complete death in less than two weeks, leaving most of the crop immature and resulting in total crop loss. The sudden appearance of wilting and death of the plants has resulted in the common name of sudden wilt for this disease.

Source of infection and spread
The precise nature of sudden wilt on cucurbits in Australia has not been fully determined. Elsewhere, in several overseas countries the fungus *Monosporascus cannonballus* has been directly implicated in the cause of the disease on melons. This fungus is soil-borne and spread by the movement of contaminated soil, by water, equipment, potted plants and vehicles. The disease is associated with fungal root infection by fungal mycelium or spores and plant stress occurring at fruit set. Several other soil-borne pathogens, especially

Fig 11.53 Sour rot of stem end on rockmelon fruit.

Fig 11.54 Collapse of rockmelon vines affected by sudden wilt during fruit fill.

Fig 11.55 Orange-brown discolouration and poor development of a root system in a plant affected by sudden wilt.

species of *Pythium* and *Fusarium*, as observed in Australia, destroy the feeder roots and invade larger roots, restricting water uptake. The high demand for water during fruit enlargement cannot be met, and the vine wilts. Fungi causing root rotting develop most rapidly in saturated or poorly aerated soil and are favoured by hot, dry weather conditions.

These conditions develop if soil structure is destroyed and soil compaction increases during bed formation. Varieties differ considerably in their susceptibility to sudden wilt, with those having vigorous root systems showing greater tolerance to the disease.

Importance

Sudden wilt is a common and serious disease of rockmelon and honeydew melon in Queensland and some parts of Western Australia. The *Monosporascus cannonballus* fungus occurs in parts of Asia, Japan, the Middle East and the USA but has not been detected in Australia.

Management

- Avoid highly susceptible varieties.

- Accurate management of irrigation is essential for sudden wilt control. Avoid over watering, particularly at fruit set. This can be a particular problem when using trickle irrigation and plastic mulch.

NEMATODES

■ ROOT-KNOT NEMATODE

Refer to the chapter on Common diseases of vegetable crops.

VIRUSES

■ CUCUMBER YELLOWS

Cause

Beet pseudoyellows virus (Crinivirus)

Symptoms

Symptoms first appear on older leaves as yellow spots followed by the development of yellow blotchy raised areas between the veins which generally remain green. The raised areas combine to form large, brittle thickened areas. Leaf margins on older leaves curl downwards. Symptoms develop progressively on the younger leaves and plants may remain stunted.

Although fruit production may be reduced, specific symptoms do not develop on fruit.

Fig 11.56 Chlorotic spotting caused by *Beet pseudoyellows virus* on glasshouse cucumber plants.

Fig 11.57 Symptoms of the *Beet pseudoyellows virus* on a mature cucumber leaf.

Fig 11.58 Distorted, chlorotic leaves of a cucurbit plant infected by *Watermelon mosaic virus*.

The symptoms of cucurbit yellows resemble those caused by poor nutrition and growing conditions, and premature aging of plants.

Symptom development is favoured by high light intensity.

Source of infection and spread

The virus is transmitted in a semi-persistent manner by the greenhouse whitefly (*Trialeurodes vaporariorum*) and has a relatively wide host range including lettuce, beet, endive, spinach and several weed species. The virus is not seed-borne and is not spread by contact.

Importance

The disease is relatively common and sometimes serious in greenhouse-grown cucumber crops in several Australian States.

Management

- Control whitefly populations in greenhouses.

- Control weeds within and around the borders of greenhouses.

- Avoid sequential plantings to reduce whitefly numbers, and remove old plantings and crop debris before planting new crops.

■ MOSAIC

Cause

Three viruses are the major cause of mosaic disease in cucurbit crops in Australia. These are the potyviruses *papaya ringspot virus* type W (PRSV-W), *watermelon mosaic virus* (WMV) and *zucchini yellow mosaic virus* (ZYMV). All belong to the potyvirus group of plant viruses.

Fig 11.59 *Papaya ringspot virus* symptoms on squash.

Fig 11.60 *Papaya ringspot virus* symptoms on a pumpkin leaf.

Symptoms

Watermelon mosaic virus causes mosaic symptoms on leaves but rarely fruit distortion. Light to dark green mottling and slight roughening of the skin can occur on fruit from affected plants.

Papaya ringspot virus (PRSV-W) causes a prominent light green and dark green mosaic pattern on leaves, which become distorted and blistered. Terminals of recently affected watermelons stand more erect, with the mosaic pattern developing later. The yield of severely affected zucchini plans reduces drastically and fruit produced is small and severely distorted. Leaves become claw-like with a severe blister mosaic pattern. Pumpkin fruit are often lumpy and distorted while watermelon fruit can develop ringspot patterns on the skin.

Zucchini yellow mosaic virus produces a severe, yellow mosaic, usually with associated leaf distortion and blistering. The first few leaves affected on rockmelons may progress from chlorosis to full necrosis. Plants are often stunted with poor fruit set. Zucchini regularly fails to set fruit, with those that do being small and distorted. Squash fruit reduce in size and may develop conspicuous yellow blotches and rings. Pumpkin fruit often develop lumps, with a yellow mottled pattern throughout.

Fig 11.61 Severe symptoms of *Papaya ringspot virus* symptoms on zucchini fruit.

Source of infection and spread

The main sources of infection for the three viruses are old, diseased cucurbit crops, cucurbits in the home garden, and weed species such as wild gherkin (*Cucumis angurin*) and paddy melon (*Citrullus lanatus*). The viruses are spread by aphids moving through or within a crop. Winged aphids may be carried several kilometres by wind. Aphids feeding on leaves for 30 seconds or less can acquire and transmit the virus.

Importance

The three viruses occur in most production areas and are a major cause of loss. Virus infection results in severe losses through reduced fruit set and fruit distortion in pumpkin, squash, rockmelon and zucchini. Papaya ringspot can severely reduce fruit set in watermelon. Cucumbers and Jap pumpkins are usually less severely affected.

Management

• Destroy old cucurbit crops as soon as harvesting is completed.

Fig 11.62 A watermelon vine infected with *Papaya ringspot virus*. Note the erect terminal growth of the plants.

Fig 11.64 A zucchini plant showing typical symptoms of virus infection.

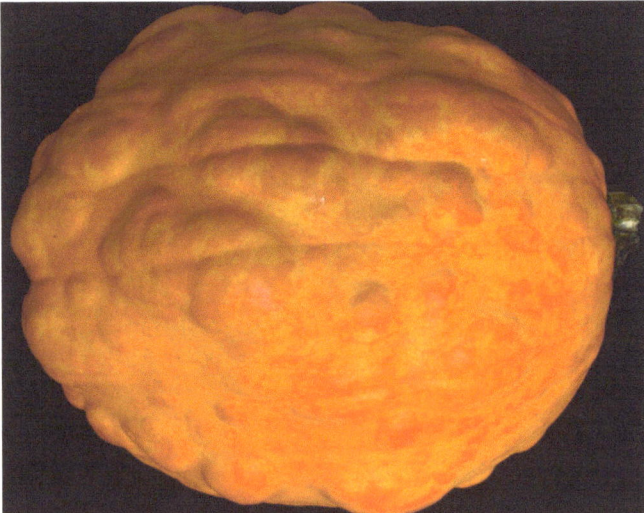

Fig 11.63 A Halloween pumpkin infected with *Papaya ringspot virus*. Fruit distortion is a characteristic crymptom.

Fig 11.65 *Zucchini yellow mosaic virus* on a pumpkin leaf.

Fig 11.66 Field symptoms of *Zucchini yellow mosaic virus*.

- Destroy weeds and volunteer cucurbits within and around crops as these harbour the viruses and/or the aphids.

- Avoid overlapping crops of cucurbits, particularly zucchini.

- Separate new crops from maturing ones as the latter are likely to have high levels of virus infection.

- Use resistant varieties, if available.

- Super-reflective plastic mulch can delay disease development in zucchini and in other cucurbit species that do not rapidly cover the mulched area. The mulch deters aphids from landing on leaves. This treatment may be used in conjunction with weekly applications of a mineral oil/insecticide spray that is applied to cover the leaf surfaces thoroughly.

- Regular insecticide applications generally have little effect on the spread of the three viruses, as feeding times necessary for transmission are so short, and a significant amount of spread occurs when aphids move through a crop, making only brief probes on leaves.

■ SQUASH MOSAIC

Cause
Squash mosaic virus (Comovirus).

Symptoms
Seedlings grown from infected seed develop green vein banding on the first or second leaf. Subsequent leaves tend to cup upwards, and produce a mottled pattern of light and dark green. Characteristic symptoms on squash plants are curved leaves, and regular projections from the veins on the leaf margins resulting from unequal growth of leaf tissue. New foliage may be symptomless or may develop yellow spots, vein clearing or leaf distortion. Infected plants appear stunted with fewer branches and fruit. Mild mottling to severe deformation may occur on fruit.

Source of infection and spread
The virus is introduced in infected seed, with seed infection levels of several per cent occurring in some honeydew melon seed-lots. *Squash mosaic virus* spreads from plant to plant on leaf-feeding beetles such as the 28-spotted ladybird.

Importance
Squash mosaic is generally a minor disease in Australia with occasional serious outbreaks resulting from sowing infected seed. The virus has been recorded from squash, pumpkin and rockmelon.

Fig 11.67 *Papaya ringspot virus* on watermelon fruit.

Fig 11.68 *Squash mosaic virus* on squash leaves.

Management

The virus is usually well controlled by eliminating seed lots containing the virus and producing seed in areas where the virus is not prevalent.

Controlling the beetle carriers (vectors) is also effective in restricting spread within a crop.

■ FURTHER INFORMATION

Babadoost M and Zitter TA (2009). Fruit rots of pumpkin. *Plant Diseases* **93**, 772–782.

Koike ST, Gladders P and Paulus AO (2007). 'Cucurbitaceae', in: *Vegetable diseases – a colour handbook*. pp. 220–251. Academic Press, New York.

Napier A and Draper V (Eds) (2009). *Pests, beneficials, diseases and disorders in cucurbits: Field identification guide*. NSW Department of Primary Industries, New South Wales.

Zitter TA, Hopkins DL and Thomas CE (Eds) (1996). *Compendium of cucurbit diseases*. APS Press, St Paul, Minnesota.

Eggplant or aubergine (*Solanum melongena*) is a member of the potato or Solanaceae family. Originating in India, the plant is widely grown in subtropical and tropical regions for its large, glossy, firm-fleshed fruit containing many fine seeds. Eggplant requires warm growing conditions and is susceptible to frost and wind damage.

Bacterial and fungal wilts can cause serious crop losses.

■ BACTERIAL WILT

Cause
The bacterium *Ralstonia solanacearum*.

Symptoms
Bacterial wilt first appears as drooping in one or more of the youngest leaves. Rapid, 'green' wilting then develops in the foliage, which is particularly noticeable during the warmest part of the day. Stunting may precede the wilting, and leaflets and leafstalks may curl downwards. In the early stages of the disease, if the stem is cut across near ground level, a light brown discolouration of the water-conducting tissues just beneath the bark is visible. This discolouration becomes darker as the disease progresses and in advanced cases, the stem cortex and pith tissues also become brown. Extensive, dark brown water-soaked lesions may also develop on the external stem, particularly at the base of the plant. For plants at an advanced stage of infection, a milky exudate is apparent

Fig 12.2 A plant infected by bacterial wilt.

Fig 12.1 Wilt and death of plants infected with the wilt bacterium *Ralstonia solanacearum*.

after several minutes if the end of a cut stem is placed in a glass of water. Below-ground symptoms are variable. Initially, root systems may appear healthy with the only evidence of infection being bacterial invasion through the root tips. Only one or a few roots may show a brown rot, but as the disease progresses, the entire root system may become discoloured and fine, feeder roots will be lost.

Source of infection and spread
The bacterium affects a very wide range of hosts among cultivated plants and weeds. The most economically important hosts are in the potato family (Solanaceae) and include potato, tomato, eggplant, tobacco and capsicum. There are five races of this pathogen. Race 1 is the principle race affecting eggplant. The pathogen infects the roots of alternative weed hosts and can also survive in the soil for extended periods in the absence of host plants.

Ralstonia solanacearum enters roots through wounds made during transplanting and cultivation, or made by nematodes or soil insects. It rapidly multiplies in the water-conducting tissues of plants, filling the tissues with bacterial cells and slime, preventing water uptake and transport, and causing the plant to wilt. Disease development occurs rapidly at high temperatures (30–35°C).

The bacterium spreads through fields in running water and via soil movement, and plant-to-plant spread occurs by root contact. Long-distance dispersal can occur through the movement of infected plants, especially seedlings.

Importance
Bacterial wilt may be a serious disease in the field during warm weather. It also is potentially severe in greenhouse production.

Management
- Avoid planting into known infested areas in summer.
- Use fumigated soil or a soil-less mix in seedling nurseries. Do not apply animal manure or excessive amounts of nitrogenous fertilisers.
- Rotations are of limited value because of the wide host range and the ability of the pathogen to survive in soil.

■ LITTLE LEAF DISEASE

Cause
Tomato big bud phytoplasma.

Symptoms
Affected plants are stunted and bushy, developing small, pale green leaves on shortened petioles. Stem internodes

Fig 12.3 Symptoms of little leaf disease in eggplant.

are shortened and plants are often bushy resulting from the leaf growth from axillary buds. Flowers are often leaf like in appearance and sterile.

Source of infection and spread
The phytoplasma has a wide host range among crop and weed hosts. The pathogen is spread from plant to plant by sap sucking leafhoppers, noteably *Orosius argentatus*. The insects frequently move into eggplant crops from adjacent weedy areas where alternative hosts are present.

Importance
The disease, although relatively common in eggplant crops, seldom reaches levels. which result in important crop losses.

Management
- Control weeds in and around crops.
- Destroy harvested crops before planting new crops nearby.

■ PHOMOPSIS FRUIT ROT

Cause
The fungus *Phomopsis vexans*.

Symptoms
Fruit spots are pale and sunken with distinct borders. The spots often originate in the calyx and expand into the fruit stalk and fruit. Affected fruit are soft and spongy and rot develops rapidly through the fruit, causing light brown discolouration of the flesh. Tiny, black fruiting bodies (pycnidia) of the fungus develop on affected tissues and may be arranged in a concentric pattern. In dry conditions, fruit shrivel and mummify. The fungus can also infect the leave and stems of eggplant.

Source of infection and spread
The fungus survives on seed and in diseased crop residues. Spores spread in water splash, on insects and on tools. The fungus can infect plants within 12 hours and symptoms develop seven to 10 days after infection. Wounds are not necessary for the pathogen to infect. Warm, wet weather favours disease development. The fungus grows well at temperatures between 21 and 32°C. The optimum temperature for infection is 29°C.

Importance
Phomopsis infection of eggplant is common in subtropical and tropical regions.

Fig 12.4 Phomopsis fruit rot.

Management

- Save seed from healthy fruit only and collect early in the season when the disease is less likely to be present.

- If the seed is not known to be disease-free, treat it in hot water (50°C for 30 minutes).

- Rotate eggplant with other crops, and ideally follow a three-year rotation out of eggplant following a disease infestation.

- Destroy crop residues promptly after harvest.

■ VERTICILLIUM WILT

Cause

The fungus *Verticillium dahliae*.

Symptoms

Plants of any age may be affected, although symptoms are more pronounced after fruit set. Early symptoms, which may

Fig 12.5 Early symptoms of Verticillium wilt.

first appear on only one side of a plant or leaf, are a slight wilting and yellowing of older leaves. As the disease develops, the leaf yellowing becomes more severe, leaf tissue dies, leaf margins curl, and leaves wither and fall from the plant. In advanced stages of the disease, plants appear stunted, with varying amounts of wilted, yellowish and dead foliage. A brown discolouration of woody stem and root tissue occurs.

Fig 12.6 Verticillium wilt. Note wilting and yellowing in older leaves.

Source of infection and spread
The fungus is a common soil inhabitant and survives in the soil for long periods as microsclerotia. It enters the roots, usually through wounds, and moves into the water-conducting tissues of the stem. The disease is favoured by cool weather. Verticillium infects other crop plants and weeds including tomato, potato, cotton and peanut. Susceptible weeds include Noogoora burr (*Xanthium pungens*), stinking roger (*Tagetes minuta*) and cobbler's pegs (*Bidens pilosa*). The fungus can also infect seed of eggplant.

Importance
Verticillium wilt is a common disease of eggplant in Australia, and may sometimes cause serious losses. The disease is also prevalent and destructivd in eggplant crops in many other countries.

Management
- Avoid planting eggplant following other susceptible crops such as tomato and potato.

- Keep crop areas free from susceptible weed hosts.

- Apply the recommended soil fumigants.

■ VIRUSES

Tomato spotted wilt virus
Refer to the chapter on Common diseases of vegetable crops.

Fig 12.7 *Tomato spotted wilt virus* symptoms on eggplant fruit.

Ginger (*Zingiber officiale*) is a perennial tropical herb in the family Zingiberaceae. Most of the Australian crop is produced in a warm, subtropical area on the Sunshine Coast of south-east Queensland. The crop prefers well-drained, sandy-loam soils and is propagated by planting 'seed pieces' cut from washed rhizomes.

BACTERIA AND PHYTOPLASMAS

■ BACTERIAL WILT

Cause

The bacterium *Ralstonia solanacearum*.

Symptoms

Biotypes (strains) 3 and 4 of *Ralstonia solanacearum* cause ginger bacterial wilt in Queensland. Biotype 3 causes a slow wilt of little significance, and biotype 4 causes rapid wilting and death. The two diseases caused by these biovars cannot be distinguished by leaf symptoms, but a distinction can be made by observing disease incidence and rate of spread.

Affected plants wilt, with yellowing in lower leaves extending upwards until all leaves are affected. The stem becomes water-soaked and readily breaks away from the rhizome. Water-conducting tissues in the stem darken.

Fig 13.2 Yellowing, rapid wilting and death are typical symptoms of bacterial wilt, biotype 4.

Fig 13.3 Bacterial wilt caused by *Ralstonia solanacearum* biotype 3. Typical symptoms include yellowing in older leaves and a slow wilt of plants.

Fig 13.1 Severe outbreak of bacterial wilt caused by biotype 4 of *Ralstonia solanacearum*.

Affected rhizomes are generally darker than normal and have water-soaked areas with pockets of milky fluid between them. When rhizomes are cut and a little pressure is applied, a milky exudate appears.

Source of infection and spread

Ralstonia solanacearum biotype 4 survives in the soil and on a wide range of crop plants and weeds including tomato, potato, capsicum, eggplant, blackberry nightshade (*Solanum nigrum*), thickhead (*Crassocephalum crepidioides*), Cape gooseberry (*Physalis peruviana*), wild gooseberry (*Physalis minima*) and wild tobacco tree (*Solanum mauritianum*). The bacterium may be introduced in infected seed pieces. It spreads rapidly in irrigation and heavy rain and in contaminated soil adhering to farm machinery.

Importance

Biotype 3, which is widely distributed in Queensland, causes only limited infection of ginger. Biotype 4, thought to have been introduced from overseas originally, in infected rhizomes, causes a rapid and severe wilt. This biotype is controlled by using disease-free planting material and by not planting infested land. Severe wilting is rare now, but is devastating when it occurs.

Management

- Do not replant infested sites.

- Plant only disease-free seed pieces. Use seed derived from tissue culture.

- Avoid rotation with susceptible crops and keep weeds under control.

- Destroy crop residues immediately after harvesting and disinfest machinery, particularly before moving to other farms.

■ BIG BUD

Cause

Tomato big bud phytoplasma.

Symptoms

Affected plants stop growing, and leaves become bunched at the top of the stem. In advanced stages of the disease, plants turn yellow and die.

Source of infection and spread

Many cultivated plants and weeds are susceptible. The organism is spread by leafhoppers, particularly *Orosius argentatus*, which breeds on weed hosts and migrates to

Fig 13.4 Big bud disease. Plants turn yellow and leaves bunch at the top of the stem.

ginger and other crop plants when weeds become poor hosts. The disease generally is more common when dry weather causes the vegetation around a crop to dry off and leafhoppers move into the crop. Subsequent spread within ginger plantings is generally very limited.

Importance

Big bud occurs sporadically, but generally is of minor importance.

Management

- Control is usually not warranted. Affected plants should be removed.

■ SOFT ROT

Refer to the chapter on Common diseases of vegetable crops.

FUNGI

■ FUSARIUM YELLOWS AND RHIZOME ROT

Cause

The fungus *Fusarium oxysporum* f.sp. *zingiberi*.

Symptoms

Affected plants are stunted and yellow. The lower leaves dry out and, eventually, the whole plant wilts and dies.

Fig 13.5 Field symptoms of Fusarium yellows are stunted plants, leaf collapse and death.

Infected rhizomes develop a brown, internal discolouration accompanied by shrivelling. In the final stages of decay, all that remains of the rhizome is the shell containing fibrous tissue. A white, cottony fungal growth may develop on the surface of stored rhizomes.

Source of infection and spread

The fungus is commonly introduced in infected seed pieces. Once introduced into soil, it may survive for many years. Infection occurs through the roots or through cracks and injuries to the skin. Pre-emergent rotting of unprotected seed pieces may occur.

Importance

This is the most serious disease of ginger in Queensland, particularly in late-harvest crops. The disease is difficult to control.

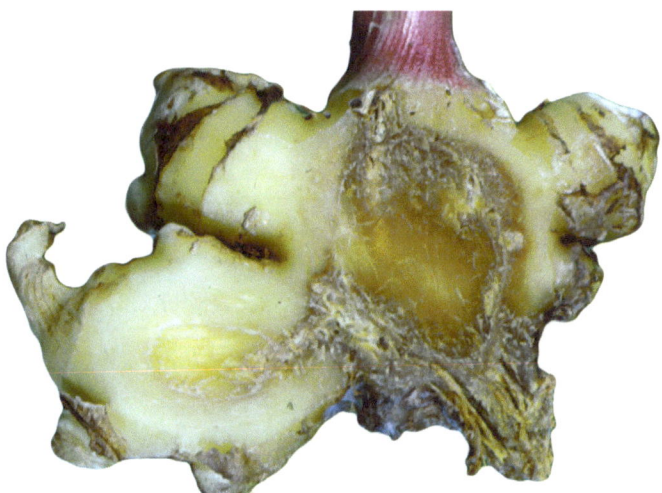

Fig 13.6 Typical rhizome symptoms of Fusarium yellows and rhizome rot.

Management

- Where possible, do not replant infested sites.

- Plant only disease-free seed pieces. During cutting, reject all rhizomes showing internal discolouration. Preferably, use planting material derived from tissue culture which has been produced under strict quarantine.

- Dip seed pieces in the recommended fungicide as soon as they are cut.

- Avoid damaging seed pieces during bagging or planting.

■ PYTHIUM RHIZOME AND SEED PIECE ROT

Cause

The oomycete *Pythium myriotylum*.

Fig 13.7 Early yellowing symptoms (centre) of Pythium rhizome and seed piece rot.

Fig 13.8 Pythium rhizome and seed piece rot showing total plant collapse in a saturated soil.

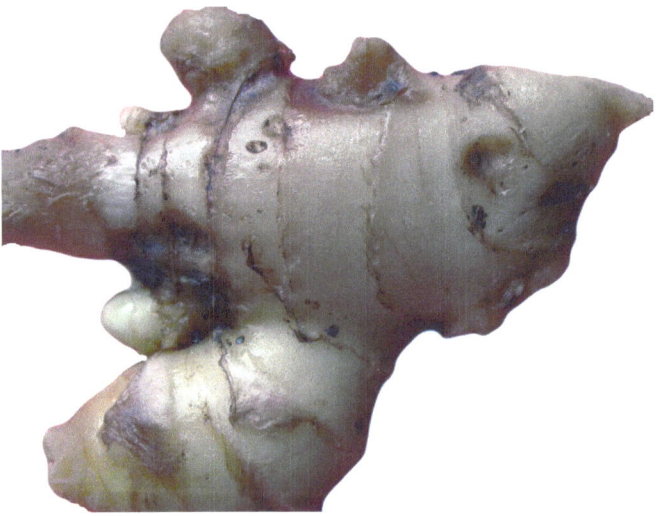

Fig 13.9 External rhizome symptoms of Pythium rhizome and seed piece rot.

Fig 13.10 Internal symptoms of Pythium rhizome and seed piece rot. Note the brown discolouration and wet rot.

Symptoms

Soft water-soaked lesions develop on roots, rhizomes and stem bases, resulting in yellowing and eventual death of the plants. Rhizomes from surviving plants are likely to develop post-harvest rot.

After harvest a brown, internal discolouration of rhizomes occurs, followed by a soft, wet rot. A white mycelial growth may also be present. Rhizomes become hollowed-out.

Source of infection and spread

The pathogen is a common soil inhabitant and wet weather and temperatures above 30°C favour disease.

Importance

The disease is severe in flooded or poorly drained soils. It can cause serious post-harvest losses when rhizomes are harvested or cut when wet or stored under moist conditions. It is most common in ginger harvested from poorly drained areas.

Management

- Select well-drained soils and plant seed derived from tissue culture that is grown in dedicated quarantine sites.

- Avoid harvesting or cutting rhizomes when wet.

- Dry seed pieces thoroughly after dipping.

- Store ginger in a cool well-ventilated place.

■ RHIZOME AND BASAL STEM ROT

Cause

The fungus *Pterula* sp.

Symptoms

Stems and leaves yellow and dry out. When plants are pulled, the stems invariably break from the rhizome.

Fig 13.11 Field symptoms of rhizome and basal stem rot.

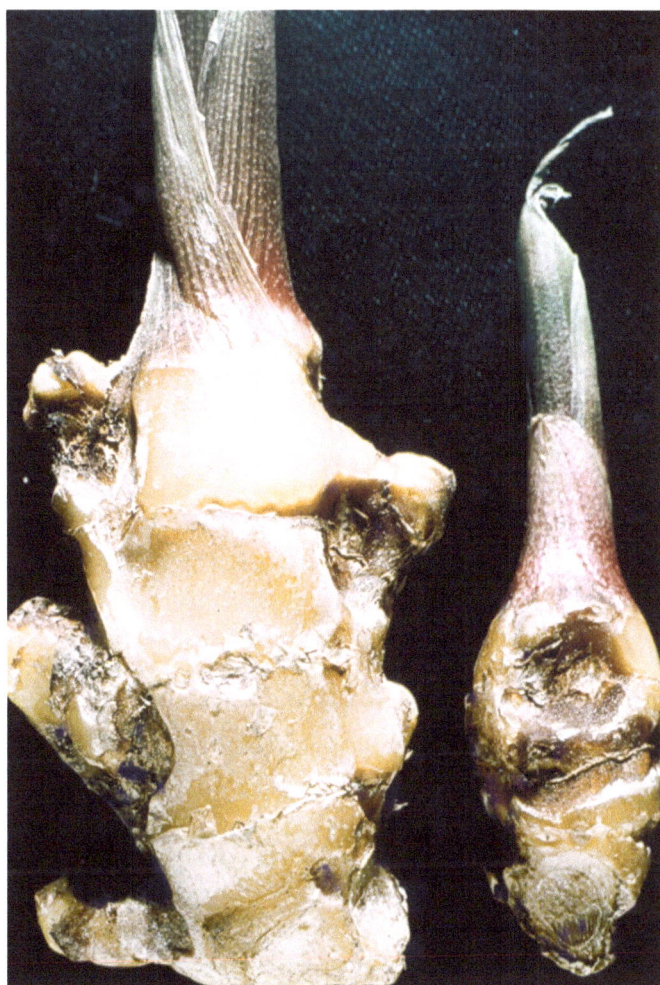

Fig 13.12 Rhizome and basal stem rot of seed piece.

Sunken spots occur beneath the scale leaves of rhizomes. Later, a dark brown rot of the rhizome occurs, and roots may be killed. In severe cases, the base of the stem hollows out and fills with a white fungal growth, which may envelop the rhizome.

Source of infection and spread
The fungus survives in the soil on undecomposed plant residues from which it sends out white strands. The most serious outbreaks of this disease have occurred after a sugarcane crop, where a large quantity of trash has remained in the soil.

Importance
The disease is prevalent but rarely serious.

Management
- Prepare land early to allow plant residues ample time to decompose.

NEMATODES

◼ ROOT-KNOT NEMATODE

Cause
The nematodes *Meloidogyne* spp.

Symptoms
Affected plants are stunted and have yellow leaves with marginal browning, and swollen, distorted roots. Rhizomes show brown, water-soaked areas in the outer tissues, particularly in the angles between shoots. Rhizomes held over for planting in the following season may show extensive breakdown in storage.

Life cycle
Refer to the chapter on Common diseases of vegetable crops.

Importance
Root-knot nematodes are a major problem in ginger, particularly in late-harvest crops.

Management
- Plant only nematode-free seed pieces. Obtain planting material from seed nurseries established from tissue culture. Clean seed may also be achieved by mulching seed-producing crops with a 10–15 cm layer of sawdust, or by heat-treating seed (48°C for 20 minutes).

- Treat soil with a registered nematicide before planting. In severe infestations, it may be necessary to treat with a nematicide after planting.

- The addition of large quantities of organic matter can reduce losses from root-knot nematode.

- Rotate with crops that are not hosts of root-knot nematode (e.g. oats and some forage sorghums).

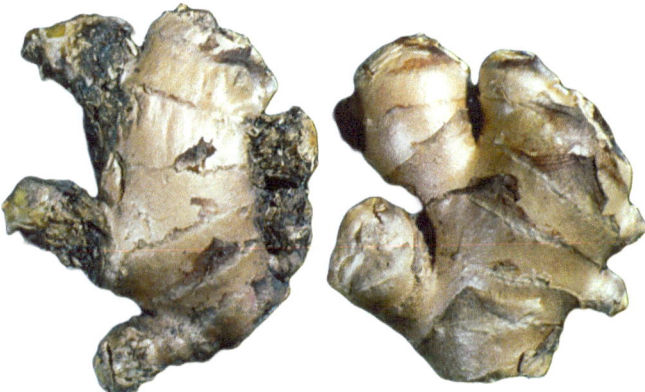

Fig 13.13 Rhizome infested with root-knot nematode (left) and a healthy rhizome (right).

Herbs and speciality crops are considered together because they are typically grown in limited quantities over small production areas or in speciality production facilities. They are produced using both field and hydroponic practices. Disease management research for these crops has been limited; plant varieties with disease resistance are often unavailable, and there are very few effective fungicides registered for their control.

ARTICHOKE

(*Cynara scolymus*, Family Asteraceae)

■ POWDERY MILDEW

Cause

The fungus *Leveillula taurica*.

Symptoms

White, powdery fungal growth, that is often difficult to distinguish from leaf hairs, develops on the underside of the older leaves. Heavily infested leaves will turn yellow and then brown and the leaves will dry out and develop a tattered appearance.

Source of infection and spread

The fungus requires a living host for survival. It persists on volunteer plants and alternate hosts including weed species. The fungus spreads as air-borne spores. Spore germination is favoured by temperatures of less than 30°C and intermediate (50–70%) to high (80–90%) relative humidity. Once an infection establishes temperatures above 30°C can accelerate both symptom development and the death of the leaf tissue. If environmental conditions favourable for powdery mildew infection persist throughout the growing season, inoculum pressure will gradually increase and so the disease will tend to be most severe at the end of the cropping period.

Importance

Powdery mildew is a common disease of artichoke in Australia and elsewhere.

Management

- Remove old crops promptly and destroy weeds and crop volunteers that may act as a source of inoculum for subsequent crops.

■ GREY MOULD

Cause

The fungus *Botrytis cinerea*.

Symptoms

The disease is primarily a post-harvest problem that can develop rapidly after infection occurs in the field. Initially, small necrotic spots form on the outer bracts of the flower head and then browning and necrosis of the inner bracts follows. A grey/brown fuzzy fungal growth develops on the bracts and on the bud apex.

Source of infection and spread

The grey mould fungus is a common invader of senescing leaves and crop debris and a pathogen of a wide range of crops and weed plants. It produces wind-borne spores that germinate and invade tissues damaged by insects, frost or mechanical means. Mild temperatures (20–25°C), free moisture and high humidity are favourable for infection. The fungus also survives in soils as hard-walled sclerotia.

Importance

Grey mould typically develops post-harvest following extended periods of mild, wet weather or heavy dews.

Management

- Apply regular applications (every seven to 10 days) of recommended protectant fungicides to minimise infection and disease spread, particularly following cool, wet weather. For best results, apply sprays prior to symptom development.

BASIL

(*Ocimum basilicum*, Family Lamiaceae)

■ BACTERIAL LEAF SPOT

Cause

The bacteria *Pseudomonas* spp.

Symptoms

Dark brown-black water-soaked patches develop rapidly on the leaves, often beginning at the leaf margins. Leaf lesions may be delimited by veins, which give them an angular shape. The hypocotyls of seedlings that are severely affected may also become water-soaked and discoloured leading to seedling collapse.

Source of infection and spread

Warm, wet weather favours the disease. Densely planted stands where air circulation is poor allow extended periods of leaf wetness that enable the disease to develop rapidly. Seedling plugs may be particularly vulnerable to infection due to moist, wet conditions during production.

Importance

Bacterial leaf spot is normally a minor disease of basil, but it can develop rapidly in warm, wet conditions.

Management

Regular applications of protectant copper fungicides should help to minimise the spread of the disease, following wet weather.

■ FUSARIUM WILT

Cause

The fungus *Fusarium oxysporum* f.sp. *basilicum*.

Fig 14.1 Bacterial leaf spot of basil causing water-soaked patches on leaves.

Symptoms

Symptoms may develop on seedlings or on mature plants. Seedlings develop black water-soaked patches on leaves and stems and a grey/brown discolouration in the internal vascular tissues and roots. Severely affected seedlings may wither and die. Mature plants are typically stunted and show characteristic foliar wilting and defoliation. A dark discolouration develops on the outer stem tissues and the lateral shoots may die back from the tips giving rise to a 'shepherd's crook' appearance. A dark grey to red/brown discolouration develops in the vascular system of the stem and in the roots. In advanced stages of the disease, the stem pith tissues may begin to break down.

Source of infection and spread

The fungus can persist in soil for years and can be spread between sites in contaminated soil. It also spreads as air-borne fungal spores produced on plant stems. Long distance dispersal occurs predominantly via contaminated seed.

Fig 14.2 Fusarium wilt on basil, showing plant wilt and stunting in an infected plant (left) compared with a healthy plant of the same age.

Fig 14.3 Fusarium wilt of basil tip dieback, giving a 'shepherds crook' appearance.

Fig 14.4 Internal stem symptoms of Fusarium wilt on basil.

Importance

Fusarium wilt is a major disease of basil in Australia, occurring in both field and hydroponic production systems.

Management

- Plant seed certified free of the disease.

- Do not collect seed from prior basil crops, particularly if these have a history of the disease.

- Avoid moving farm machinery or soil between infected and non-infected sites.

- In hydroponic systems or nurseries, thoroughly disinfect all pots and growing trays with an agricultural sanitiser

product before planting, particularly when there has been a history of the disease.

■ CHARCOAL ROT

Cause

The fungus *Macrophomina phaseolina*.

Symptoms

Plant stems develop a brown discolouration that may extend from the shoot tips into the plant crowns and roots. The affected stems defoliate and tiny black pimple-like fruiting bodies of the fungus (pycnidia) develop within the infected stem tissues. The outer root tissues may rot away, leaving only the rotten inner root steles remaining.

Source of infection and spread

The disease is not well understood, although infection processes are likely to be similar to those of other diseases caused by this pathogen. On other crops, disease development is favoured by hot, dry weather. The fungus can survive in soils for long periods and infection may occur following movement of infested soil. It may also persist in crop debris. Although *Macrophomina* is a seed-borne pathogen of many crops, it is not known if the fungus is transmitted in basil seed.

Importance

Charcoal rot is a sporadic but emerging disease in basil field production in Australia.

Management

- Deep-plough old crops and ensure residues have completely broken down before planting a subsequent crop.

Fig 14.5 Charcoal rot of basil.

■ BLACK SPOT

Cause

The fungus *Colletotrichum gloeosporioides*.

Symptoms

Circular or irregular necrotic spots develop on leaves. The spots frequently start at the leaf margins and then coalesce to give the appearance of a marginal leaf necrosis. Older leaves appear to be more susceptible to infection.

Source of infection and spread

Diseases caused by *Colletotrichum* spp. are favoured by high relative humidity and warm temperatures that lead to extended periods of leaf wetness. In very dense crops, leaves deep in the canopy may remain wet for extended periods, allowing symptoms to develop.

Importance

Black spot is a minor disease in Australian basil crops. In Italy, the same pathogen is reported to cause lesions that girdle plant stems, killing the plants. This symptom has not been found in Australian basil crops.

Management

• Although not usually required, regular application of protectant copper fungicides should help to minimise the spread of the disease, following wet weather.

Fig 14.6 Foliar symptoms of black spot on basil.

Fig 14.7 *Tomato spotted wilt virus* on basil.

■ TOMATO SPOTTED WILT

Cause

Tomato spotted wilt virus (Tospovirus).

Refer to the chapter on Common diseases of vegetable crops.

CORIANDER

(*Coriandrum sativum*, Family Apiaceae)

■ BACTERIAL LEAF SPOT

Cause

The bacterium *Pseudomonas syringae* pv. *coriandricola*.

Symptoms

Angular, water-soaked lesions develop on leaves, often beginning at the leaf edges. Lesions are light brown, with a darker red-brown margin and delimited by the leaf veins. The lesions are apparent on both sides of the leaves and may coalesce to give the leaves a blighted appearance when the infection is severe.

Source of infection and spread

Warm, wet weather favours the disease. Extended periods of leaf wetness allow the bacteria to multiply causing rapid symptom development. The bacterium is seed-borne.

Importance

Bacterial leaf spot can be a major problem in coriander crops in Australia and elsewhere.

Fig 14.8 Coriander leaves affected by bacterial leaf spot.

Fig 14.9 Bacterial leaf spot of coriander.

Management

- Plant only high quality seed. Seed suspected of contamination should not be planted.

- Regular applications (every seven to 10 days) of copper-based fungicides should help to minimise infection and disease spread. For best results, apply sprays prior to symptom development.

- In hydroponic systems or nurseries, thoroughly disinfect all pots and growing trays with an agricultural sanitiser product before planting, particularly when there has been a history of the disease.

DILL

(*Anethum graveolans*, Family Apiaceae)

■ POWDERY MILDEW

Cause
The fungus *Oidium* sp.

Symptoms
White powdery fungal growth appears on the leaves, stems and flower stalks. When infection is severe, twisting and malformation of the foliage may occur and the leaves may die prematurely.

Source of infection and spread
The fungus has a wide host range, affecting many plants within the Apiaceae family. Alternative hosts may be an important source of inoculum. The spores are wind-borne and may be carried long distances in the air. High humidity and moderate temperatures favour infection and disease development. Spores may be produced seven to 14 days after infection, when conditions are favourable for disease.

Importance
Powdery mildew is the most common disease of dill in Australia.

Management
- Apply regular applications (every seven to 10 days) of sulphur-based fungicides to minimise infection and disease spread. For best results, apply sprays prior to symptom development.

- Destroy old crops promptly.

- Rotate dill with crops that do not belong to the Apiaceae family.

Fig 14.10 Close view of pustules of the rust fungus *Puccinia hieracii* on the underside of an endive leaf.

Fig 14.11 Tangle top on lemongrass. Note the tangled, 'looped' arrangement of emerging leaves.

ENDIVE

(*Cichorium endiva*, Family Asteraceae)

■ RUST

Cause

The fungus *Puccinia hieracii*.

Symptoms

Dark brown to red-brown pustules form on both sides of the leaves. Severe infection can cause leaf chlorosis and decline.

Source of infection and spread

The fungus spreads long distances as wind-borne spores produced from the leaf pustules. The disease can persist on endive and chicory crops, as well as other members of the Asteraceae.

Importance

Rust is the most common disease of endive in Australia.

Management

- Destroy infected crops promptly after the final harvest and before replanting to minimise disease spread to subsequent crops.

- Control Asteraceae weeds in and around endive crops, particularly when there has been a history of disease.

LEMONGRASS

(*Cymbopogon citratus*, Family Poaceae)

■ TANGLE TOP

Cause

The fungus *Myriogenospora atramentosa*.

Symptoms

Infected plants may be stunted and produce smaller tillers than usual. The young, newly emerging leaves develop a tangled 'looped' arrangement when infection by the fungus causes the tips of the unfolding leaves to adhere to the bases of adjacent leaves.

Source of infection and spread

The fungus spreads in vegetative propagation material. It overwinters as hyphae in dormant buds, and infects newly emerging leaves the following year. Fungal spores also develop, and warm, wet weather favours infection.

Importance

Tangle top is a very minor disease of lemongrass in Australia.

Management

- Only plant asymptomatic material.

- Do not plant material derived from stock plants showing symptoms of the disease.

- In crops with a history of the disease, spray with regular applications of recommended protectant fungicides, particularly following periods of warm, wet weather.

■ RUST

Cause

The fungi *Puccinia cymbopogonis* and *Puccinia nakanishikii*.

Fig 14.12 Lemongrass rust.

Fig 14.13 Powdery mildew on mint. Note the white fungal growth on the underside of leaves.

Symptoms

Elongated, dark brown lesions develop on both sides of the leaves. Lesions coalesce to form a general leaf blight and leaf death. Brown pustules containing spores of the fungus develop on the lower leaf surfaces.

Source of infection and spread

The life cycle of this disease is not well understood, although infection presumably occurs as wind-borne spores. Co-infection of crops with both species of the rust pathogen is possible.

Importance

Rust is a common disease of lemongrass in Australia.

Management

- Regular applications of recommended protectant fungicides may help to limit infection.

MINT

(*Mentha* spp., Family Lamiaceae)

■ POWDERY MILDEW

Cause

The fungus *Golovinomyces biocellatus*.

Symptoms

White, powdery fungal growth develops on both upper and lower leaf surfaces. Leaves eventually turn yellow and drop when infection is severe.

Source of infection and spread

The fungus requires a living host for survival. It is able to persist on alternative hosts in the Lamiaceae family as well as on mint crop debris, and spreads between hosts as wind-blown spores.

Importance

Powdery mildew is a sporadic but widespread disease of mint in Australia and elsewhere.

Management

- Plough in old crops soon after harvest to limit the spread of the pathogen to subsequent plantings. Ensure residues have completely broken down before planting a subsequent crop.

- Regular applications (every seven to 10 days) of sulphur-based fungicides should help to minimise infection and disease spread. For best results, apply sprays prior to symptom development.

■ RUST

Cause

The fungus *Puccinia menthae*.

Symptoms

Discrete, dark brown pustules form on both sides of infected leaves. Severely infected leaves turn yellow and defoliate.

Source of infection and spread

The rust fungus has a complicated life cycle involving five spore stages produced on living hosts. The brown urediniospores are associated with the destructive stage of the disease. In Australia, on some hosts the fungus is

Fig 14.14 Defoliation in mint plants, caused by a severe rust infection. Brown pustules of the fungus develop on infected leaves.

able to survive throughout the year as urediniospores, whereas the full life cycle involving all the spore stages occurs on other hosts. The fungus persists on alternative hosts in the Lamiaceae family as well as infected mint crops. The urediniospores disperse over large distances in wind and infection is favoured when temperatures range between 10–25°C and leaves are wet for at least six hours.

Importance
Rust is the most important disease of mint in Australia.

Management
- Crops should be monitored for rust incidence and severity.

- Spray the recommended systemic fungicide. Research conducted in Victoria indicates that the disease can be eradicated if fungicide is applied when rust pustules occur on less than 60% of leaves.

- At the end of the growing season, spray the crop with a contact desiccant herbicide to help minimise the amount of inoculum that is available to re-infect the crop in the following season.

■ FURTHER INFORMATION

Edwards, J (1999). *Control of mint rust on peppermint – epidemiology and chemical control*. Rural Industries and Development Corporation, Canberra, ACT. *Mushroom.*

MUSHROOM

(Agaricus bisporus)

■ BACTERIAL BLOTCH (BROWN BLOTCH)

Cause
The bacterium *Pseudomonas fluorescens* biotype G.

Symptoms
Light brown superficial lesions develop initially on the caps. When excess moisture is present, the initial pale blotches turn dark brown and spread over the entire surface of the mushroom cap. The stems may also become infected. Symptoms can also manifest post-harvest, particularly when mushrooms are stored at fluctuating temperatures in which condensation forms on their surfaces.

Source of infection and spread
The pathogen is generally present wherever mushrooms are grown, but, as an opportunistic organism, it only causes disease when environmental conditions are favourable. Wet conditions and temperatures in excess of 20°C favour the disease. Symptoms often develop after watering, when the caps are not dry within two to three hours.

Importance
Bacterial blotch is a common and potentially devastating disease of mushrooms in Australia and elsewhere.

Management
- Regulate the temperature and relative humidity to ensure that condensation dries from the caps within two to three hours.

- Ensure there is sufficient ventilation, particularly after watering.

- Apply a chlorine solution (150 ppm) each time the casing is watered.

■ WET BUBBLE

Cause
The fungus *Mycogone perniciosa*.

Symptoms
Infected mushrooms become misshapen, with swollen stems and caps. A light brown discolouration develops on their surfaces and red/brown droplets of liquid may ooze from the tissues. A characteristic odour is produced in the mushroom house when this pathogen is present.

Source of infection and spread
The fungus is spread as spores and mycelium, which initially infect casing and then are transferred throughout the mushroom house by phorid and sciarid flies, in dust, on workers and on infected picking trays and equipment. Water run-off is also an important mechanism of spread.

Fig 14.15 Symptoms of wet bubble disease caused by *Mycogone perniciosa*.

Fig 14.16 Dry bubble symptoms in mushroom.

Importance

Wet bubble is a serious disease of mushrooms in Australia and elsewhere.

Management

- Good hygiene and early recognition of the disease is essential for control.

- Control phorid and sciarid flies in the mushroom house.

- Incorporate the recommended fungicides into the peat prior to casing. Alternatively, apply a fungicide spray to the surface of the casing at the earliest sign of the disease.

- Ensure the mushroom house is properly 'cooked out' at the end of the production period.

■ DRY BUBBLE

Cause

The fungus *Verticillium fungicola*.

Symptoms

Infected mushrooms become discoloured, cracked or shrivelled, and develop crooked or swollen stems in the later stages of the disease. Mushrooms affected by dry bubble do not ooze droplets of liquid and no odour is associated with infection, which distinguishes this disease from wet bubble.

Source of infection and spread

The fungus produces spores that are easily spread throughout the mushroom house in water and in fine air-borne water droplets. Initial infection often occurs at casing time and then secondary spread of spores occurs

via fly vectors, in dust, on workers, in water droplets produced during high-pressure water cleaning, and on mice that invade the mushroom house. Warm, moist environmental conditions are optimal for infection of the developing fruiting bodies as they emerge.

Importance

Dry bubble is a very serious disease of mushrooms. Strains of the pathogen with resistance to the fungicide carbendazim and prochloraz have been reported around the world.

Management

- Good hygiene and early recognition of the disease is essential for control.

Fig 14.17 Dry bubble disease showing cracked and shrivelled mushrooms.

- Control mites, flies and vermin in the mushroom house.

- Incorporate the recommended fungicides into the peat prior to casing. Alternatively, apply a fungicide spray to the surface of the casing at the earliest sign of the disease.

- Ensure the mushroom house is properly 'cooked out' at the end of the production period.

- Avoid the use of high-pressure water hoses for cleaning the production facility. Instead, use low pressure or a 'squeegee'.

- Monitor for disease symptoms each day and cover any infected mushrooms with table or rock salt, or spray them with 80% alcohol at the earliest sign of symptom development.

◼ COBWEB DISEASE

Cause
The fungus *Hypomyces rosellus*.

Symptoms
A white web of mould grows rapidly across the mushroom beds, covering the mushrooms and surrounding casing and causing a rapid decay. Sometimes, the mycelium and infected mushrooms develop an orange to red discolouration. Light to brown spotting forms on the mushroom tissues and affected mushrooms eventually die.

Source of infection and spread
The fungus is a soil inhabitant, and probably enters the mushroom house with soil dust. Once present, it produces abundant spores which are spread rapidly by air movement, water splash and excess water run-off. Warm, moist or humid conditions favour infection. Higher casing moisture levels and low rates of evaporation provide conditions conducive to disease development. Typically, cobweb disease appears only on the later flushes.

Importance
Cobweb disease is a significant problem in Australia and elsewhere. Strains of the pathogen with decreased sensitivity to ther fungicide prochloraz have been reported.

Management
- Good hygiene and early recognition of the disease is essential for control.

- Ensure that no soil contaminates the compost or casing.

- If disease develops, apply a fungicide spray to the surface of the casing immediately after harvest of the first flush.

- Monitor for disease symptoms each day, and cover any infected mushrooms with alcohol-drenched paper towel and then cover this with table salt or rock salt.

◼ VIRUSES

Cause
At least six viruses have been detected in mushroom crops in Australia.

Symptoms
A wide range of symptoms may be associated with virus infection. Reduced production is the most consistent indication of probable virus infection. Bare patches in beds

Fig 14.18 Cobweb disease caused by *Hypomyces rosellus*.

Fig 14.19 Virus symptoms in mushroom.

resulting from slow mycelial growth are often an indication of viral infection. Infected mushrooms may develop splitting, bending and swelling of the stalk. Long stalks and small caps giving a 'drum-stick' effect, and a premature opening of the veil is another indication of possible viral infection.

Infections may be symptomless and several different viruses are commonly involved in a virus disease outbreak.

Source of infection and spread

Viruses carried in spores from infected mushrooms are a common means of virus dispersal. Infected mycelium can also survive in mushroom trays and infect new crops. Infection at spawning often leads to high disease levels while infection entering at casing and cropping generally results in lower levels of disease.

Importance

Virus infection can be a serious cause of loss or crop failure in mushroom production.

Management

Strict hygiene is essential to prevent or eliminate infection.

- Disinfect all machines used for filling, spawning and casing; and working areas and growing rooms after each operation.

- Plan work areas to provide one-way movement from the clean to the dirty areas during normal operations. Filter the air entering spawn-running rooms, growing rooms and work areas.

- Sterilise mushroom trays after cropping and dispose of spent compost.

■ FURTHER INFORMATION

European and Mediterranean Plant Protection Organization (2000). *Mushrooms. Guidelines on good plant protection practice.* EPPO Standards. European and Mediterranean Plant Protection Organization, Paris. http://archives.eppo.org/index.htm/PP2_GPP/pp2-20-e.doc

Fletcher JT and Gaze RH (2008). *Mushroom pest and disease control—a colour handbook.* Academic Press/Manson Publishing, London.

Nair NG (1984). *Diseases of cultivated mushrooms.* Agfacts H8.AB.30. Department of Agriculture, New South Wales.

OREGANO

(*Origanum marjorana/Origanum vulgare*, Family Lamiaceae)

Fig 14.20 Uneven growth in parsley caused by root-knot nematode.

■ ROOT-KNOT NEMATODE

Cause

The nematodes *Meloidogyne* spp.

Refer to the chapter on Common diseases of vegetable crops.

PARSLEY

(*Petroselinum crispum*, Family Apiaceae)

■ ALTERNARIA LEAF BLIGHT (LEAF SCORCH)

Cause

The fungus *Alternaria petroselini*.

Symptoms

Brown to black spots develop on leaves, initially at the leaf margins. The spots expand until the entire leaf turns yellow then brown and collapses. Complete defoliation of

Fig 14.21 Alternaria leaf blight symptoms in curl-leaf parsley.

plants can occur. Very young and very old leaves are more severely affected. A black necrosis of the crown and upper part of the taproot can also occur, resulting in collapse of the plant. Seed-borne infection can result in seedling plants damping-off.

Source of infection and spread

Seed-borne infection often initiates the disease within a crop. Disease development is favoured by warm (28°C is optimal), wet, humid weather. Fungal spores disperse in wind, water splash and contact between plants.

Importance

Alternaria leaf blight is a common disease of parsley worldwide and can cause severe crop losses when weather conditions favour disease development.

Fig 14.22 Alternaria leaf blight symptoms in flat-leaf parsley.

Management

- Use high quality seed, preferably tested for freedom from seed-borne infection.

- Hot-water treat seed to remove or reduce seed-borne infection (50°C for 20 minutes).

- Avoid planting new crops next to older, infected plantings.

- Destroy old crops promptly to minimise spread of the fungus to new plantings.

- Minimise leaf wetness by using trickle irrigation instead of overhead irrigation.

■ ROOT ROT

Cause

The oomycetes *Phytophthora* and *Pythium*.

Fig 14.23 Leaf yellowing and collapse of field-grown parsley affected by root rot.

Fig 14.24 Root rot symptoms in hydroponic parsley seedlings infected by *Pythium*.

Fig 14.25 Severe root rot of a field-grown parsley plant.

Symptoms

Both pathogens cause pre- and post-emergence damping-off in seedlings and the death and decline of mature plants. The above-ground symptoms are a rapid wilt of foliage, a rot of the leaf stalk bases and collapse of shoots. Plants die or remain stunted and yellow.

Roots are dull brown and spongy. *Pythium* tends to attack the tips of the lateral roots and infects at the crown or upper root, destroying most of the lateral root system. Infection by *Phytophthora* usually begins at the root tips and travels up the roots. Infected roots are light to dark brown in colour with lateral roots still present in many cases.

Source of infection and spread

The pathogens are common soil inhabitants and infect a very wide range of plants. They infect roots via motile zoospores and are favoured by wet, poorly drained soils.

Fig 14.26 Poor lateral root development in a parsley plant infected with *Pythium* species.

Importance

Root rots caused by *Pythium* and *Phytophthora* are frequently serious in southern Australian production areas, during late autumn and winter following cold wet weather when temperatures are below 10°C.

Management

- Plant on raised beds to improve drainage and reduce water-logging.

- Time irrigation to avoid wet or dry extremes of soil water.

- Avoid planting low-lying areas.

- Incorporate crop debris into soil to encourage breakdown as *Pythium* and *Phytophthora* can survive saprophytically.

- Apply the recommended fungicides.

■ SCLEROTINIA ROT

Refer to the chapter on Common diseases of vegetable crops.

■ SEPTORIA LEAF SPOT

Cause

The fungus *Septoria petroselini*.

Symptoms

The disease appears as small, tan leaf spots surrounded by a dark reddish-brown margin. Dark fruiting bodies of the fungus are often present in the spots. As the disease progresses, the spots become more numerous, coalesce and cause yellowing and blighting of the foliage.

Fig 14.27 Septoria leaf spot on parsley.

Source of infection and spread

The fungus can be seed-borne. Spores of the fungus spread in wind-driven rain and overhead irrigation. The spores also spread on machinery and workers in contact with wet foliage. The disease is favoured by mild temperatures (20–25°C), and high humidity from, for example, morning dews or overhead irrigation.

Importance

Septoria leaf spot is the most common foliage disease of parsley in Australia. It affects both flat and curled leaf varieties.

Management

- Use high quality seed.
- Flat-leaf varieties are generally more susceptible to *Septoria* leaf spot than curly-leafed types.
- Use drip or trickle irrigation rather than overhead sprinkler systems to reduce leaf wetness and humidity.
- Rotate parsley with other crops.
- Apply recommended fungicides.

■ STEMPHYLIUM LEAF SPOT

Cause

The fungus *Stemphylium* sp.

Symptoms

Initially, small chlorotic lesions appear on the petioles and leaves. In time, these develop into brown, necrotic circular to irregular spots that may coalesce to produce a leaf blight if infection is severe.

Source of infection and spread

The disease is favoured by warm, wet weather. Extended periods of leaf wetness allow the fungus to produce masses of spores that spread in wind and water-splash.

Importance

Stemphylium leaf spot is a minor disease of Australian parsley crops.

■ ROOT-KNOT NEMATODE

Cause

The nematodes *Meloidogyne hapla* and *M. incognita*.

Symptoms

Symptoms include stunted growth, small plants, leaf yellowing and wilting during the warmer part of the day. Affected plants generally occur in patches with the number and size of the patches increasing as the season progresses. Secondary bacterial soft rot often develops in parsley roots infected initially by root-knot nematodes.

Importance

Root-knot nematodes are an important cause of decline in parsley grown on lighter sandy soils in Queensland and New South Wales production areas.

Refer to the chapter on Common diseases of vegetable crops.

Fig 14.28 Stemphylium leaf spot on Italian parsley.

Fig 14.29 Root-knot nematode causing severe galling on infected parsley roots.

Fig 14.30 Field symptoms of root-knot nematode in a flat-leaved parsley crop.

■ APIUM VIRUS

Cause
Apium virus Y (Potyvirus).

Symptoms
On young leaves, the virus causes vein clearing and a yellow to light green interveinal mottling. Narrow, twisted mottled leaflets occur on mature foliage.

Source of infection and spread
Many aphid species are able to transmit the virus and require only very short feeding times to complete the process. The virus does not persist in the aphid. The host range of the virus is restricted to several Apiaceae species, including sea celery (*Apium prostratum*) and hemlock (*Conium maculatum*).

Fig 14.31 Leaf mottling and chlorosis (left) caused by *Apium virus Y*.

Management
- Avoid planting new crops adjacent to older parsley crops.
- Implement a host-free period of several months if the virus is causing serious damage.
- Systemic insecticides are unlikely to prevent virus spread because of the very short transmission periods.

ROCKET

(*Eruca vesicaria* ssp. *sativa*, Family Brassicaceae)

■ BACTERIAL BLIGHT

Cause
The bacterium *Pseudomonas syringae*. In the USA, *P. syringae* pv. *alisalensis* causes a bacterial blight on rocket that produces symptoms very similar to those seen on rocket in Australia. Identification of the Australian pathovar is required to determine if these two diseases are the same.

Fig 14.32 Bacterial blight of rocket.

Fig 14.33 Bacterial blight of rocket causing angular water-soaked lesions on leaves.

Symptoms

Dark, angular, water-soaked lesions develop on leaves, predominantly along the leaf veins. The lesions may be surrounded by a yellow halo. With age, the centres of the lesions dry out, leaving light brown desiccated tissues. The lesions are apparent on both sides of the leaves.

Source of infection and spread

The disease is favoured by warm, wet weather. Extended periods of leaf wetness allow the bacteria to multiply causing rapid symptom development.

Importance

Bacterial blight is a sporadic but potentially severe disease of Australian rocket crops.

Fig 14.34 White blister caused by the fungus *Albugo candida* on rocket.

Management

- Eliminate the use of overhead sprinkler irrigation.
- Regular applications (every seven to 10 days) of copper-based fungicides may help to minimise infection and disease spread. For best results, apply sprays prior to symptom development.
- In hydroponic systems or nurseries, thoroughly disinfect all pots and growing trays with an agricultural sanitiser product before planting, particularly when there has been a history of the disease.

■ WHITE BLISTER (WHITE RUST)

Cause

The oomycete *Albugo candida*.

Refer to the description of white blister in the chapter on brassicas.

SPINACH

(*Spinacia oleracea*, Family Amaranthaceae)

■ ANTHRACNOSE

Cause

The fungus *Colletotrichum dematium*.

Symptoms

Small water-soaked circular spots develop on leaves. The spots enlarge, become chlorotic and later turn brown or tan in colour. The spots may become dry and papery in texture and merge into large dead areas causing blighting of the foliage. Tiny black fruiting bodies (acervuli) of the fungus form on diseased tissue and are visible with the aid of a hand lens.

Source of infection and spread

The fungus survives in infected crop residues and is carried on spinach seed. Fungal spores spread in water splash from rain or overhead irrigation. Wet weather, high humidity and limited air movement in dense leaf canopies favour disease development.

Importance

The disease occurs in most countries where spinach is grown, and first found in several Australian states in 2004.

The disease may be serious following wet weather, causing stunting and quality defects in both bunching and baby spinach crops.

Management

- Reduce periods of leaf wetness by using overhead irrigation early in the day to allow leaves to dry.

- Avoid growing successive crops of spinach on the same ground to reduce inoculum levels.

■ DOWNY MILDEW (BLUE MOULD)

Cause

The oomycete *Peronospora farinosa* f.sp. *spinaciae*.

Symptoms

Initial symptoms are indefinite yellowish areas on the upper leaf surface with a corresponding grey to purple growth of the pathogen on the underside of the leaf. The affected areas later become bright yellow. As the disease develops the affected areas enlarge and leaves become curled, distorted and may completely rot.

Source of infection and spread

The pathogen survives in living plants, crop residues and in spinach seed. Spores dispersed in wind and water splash. The dense crop canopy of spinach creates an ideal environment for rapid disease development, particularly under cool, wet conditions.

Importance

Downy mildew is one of the most widespread and destructive diseases of spinach worldwide. The pathogen is genetically diverse, occurring as a series of at least 10 races worldwide.

Fig 14.35 Downy mildew on spinach leaves.

Management

- Grow resistant varieties. Select varieties according to the races of the pathogen present in the region. A single gene confers resistance and races of the pathogen that can overcome a particular resistance gene, may occur reasonably frequently.

- Apply recommended fungicides to seed or seedling plants to control the disease in young plants.

WATERCRESS

(*Rorippa nasturtium-aquaticum*, Family Brassicaceae)

■ CERCOSPORA LEAF SPOT

Cause

The fungus *Cercospora nasturtii*.

Symptoms

Circular to oval light brown lesions form on the leaves. The lesions may coalesce to give a blight-like symptom.

Source of infection and spread

The fungus produces spores that are both air-borne and dispersed via splashing water.

Importance

Leaf spot is a common disease of watercress both in Australia and overseas.

Management

- Destroy infected crops promptly after the final harvest and before replanting to minimise disease spread to subsequent crops.

- Regular applications (every seven to 10 days) of the recommended protectant fungicide may help to minimise infection and disease spread. For best results, apply sprays prior to symptom development.

■ SCLEROTINIA ROT

Cause

The fungus *Sclerotinia sclerotiorum*.

Refer to the chapter on Common diseases of vegetable crops.

Lettuce (*Lactuca sativa*) belongs to the Asteraceae family and is closely related to the wild or prickly lettuce (*Lactuca serriola*). Lettuce is an annual plant of Mediterranean origin, which is now grown worldwide as a fresh salad ingredient. It prefers a relatively cool climate and suffers from several disorders when grown under high temperatures.

The types grown include crisphead or iceberg, romaine or cos, butterhead, and leaf lettuce including oak leaf, coral and monet varieties.

A considerable proportion of the Australian crop is grown hydroponically.

Important disease problems include bacterial leaf spots, Sclerotinia drop, downy mildew and *Tomato spotted wilt virus*.

BACTERIA

■ BACTERIAL LEAF SPOT AND HEAD ROT (DRY LEAF SPOT)

Cause
The bacterium *Xanthomonas campestris* pv. *vitians*.

Symptoms
Small (2–5 mm), translucent, irregular or angular water-soaked spots develop on leaves. These later become dark brown and papery. If a large number of spots occur on a leaf, affected areas become yellow, then brown and die. Older leaves are more susceptible. Hearts may rot, or plants may not develop hearts if infected when young.

Source of infection and spread
The bacterium is seed-borne, although commercial lettuce seed is infested very infrequently. Disease development is favoured by cool, showery weather. The bacteria spread in wind-blown water droplets, water splash and in overhead irrigation. It can also survive on undecomposed lettuce residues in the soil, and epiphytically on weed hosts.

Importance
Bacterial leaf spot is a common disease, with severe outbreaks often occurring following cool, wet weather.

Management
* A regular protectant copper spray program will slow disease development.
* Grow resistant or tolerant varieties.

■ BACTERIAL ROTS AND VARNISH SPOT

Cause
The bacteria *Pseudomonas cichorii*, *P. marginalis* and *P. viridiflava*.

Fig 15.1 Bacterial leaf spot on lettuce.

Fig 15.2 Varnish spot on lettuce.

Fig 15.3 Leaf spotting on lettuce caused by *Pseudomonas*.

Symptoms

Varnish spot is usually caused by *Pseudomonas cichorii*. Symptoms develop as heads mature, with a shiny brown discolouration occurring on the inner leaves and the larger leaf veins turning orange to brown. Affected areas remain firm and do not decay. Other bacteria, particularly *P. marginalis* and *P. viridiflava*, cause leaf spotting and leaf blight, vascular blackening and soft rotting, particularly during warm to hot weather.

Source of infection and spread

The diseases can occur in lettuce grown in the field or in hydroponic systems. Varnish spot is common in hydroponic systems and in water supplies contaminated with *P. cichorii*. The bacteria survive in other hosts, in soil and in crop debris. Bacteria spread in water splash. Cool temperatures favour varnish spot development, and damaged or injured plants are more likely to be infected. Leaf spotting is common during cooler temperatures, with soft rots rapidly developing during warm weather.

Importance

The diseases are common and sometimes serious in both hydroponic systems and field plantings.

Management

- Use the recommended chemicals to control bacterial pathogens in hydroponic systems.

- Use furrow or trickle irrigation rather than overhead sprinkler systems in field-grown crops. If overhead irrigation is used, schedule irrigations to prevent watering during the evening and night.

■ CORKY ROOT

Cause

The bacterium *Rhizomonas suberifaciens*.

Symptoms

Plants lack vigour and remain small and stunted. Roots show a yellow-brown banded discolouration that becomes dark brown, raised and roughened (corky) in appearance and brittle to touch. Eventually, very little root system remains.

Source of infection and spread

The bacterium builds up in soil used for continuous lettuce production. The common sowthistle (*Sonchus oleraceus*) is also a host of the pathogen, and when infected, it develops symptoms similar to those in the early stages of corky root in lettuce. The disease is more of a problem in warm soils treated with high rates of nitrogen fertilisers.

Importance

Corky root has been found in the Lockyer Valley in south-east Queensland. There is limited information about its distribution in other Australian lettuce production areas.

Management

- Plant lettuce varieties that are resistant to corky root.

- Avoid consecutive lettuce crops, and instead, grow lettuce in rotation with non-host plants.

- Trickle irrigation may reduce corky root severity.

Fig 15.4 Corky root symptoms on lettuce roots.

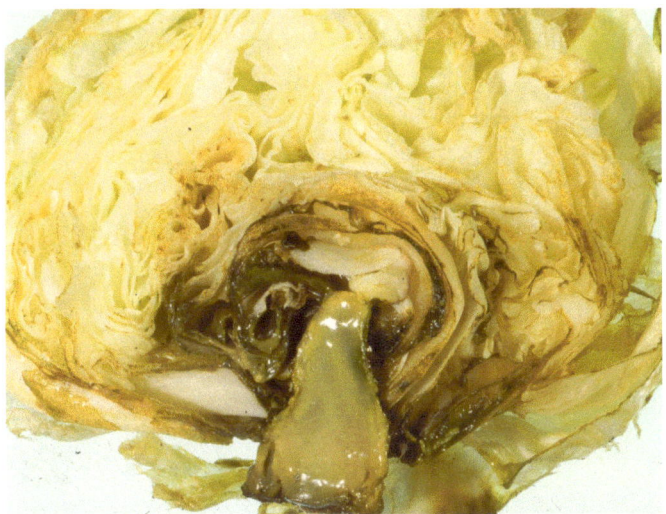

Fig 15.5 Bacterial soft rot.

Fig 15.6 Anthracnose symptoms.

- Use lettuce transplants rather than direct seeding crops. Corky root is less severe in the fibrous root systems of transplants compared with the central taproots of direct-seeded plants.

■ SOFT ROT

Cause
The bacterium *Erwinia carotovora*.

Refer to the chapter on Common diseases of vegetable crops.

FUNGI

■ ANTHRACNOSE

Cause
The fungus *Microdochium panattonianum*.

Symptoms
Small (2–3 mm) water-soaked spots develop on the lower leaves of infected plants. These enlarge to straw-coloured, circular to angular spots up to 4 mm in diameter. The centres of the spots eventually fall out, giving a characteristic shot-hole effect. Sunken, elliptical spots also develop on the leaf midrib. If the disease is severe the lesions will coalesce and cause significant leaf dieback and plant stunting.

Source of infection and spread
The fungus can survive in soil for several years as microsclerotia, and also survives on infected crop debris. Prickly lettuce (*Lactuca serriola*) is also a host. Spores

spread in rain or overhead irrigation. Cool, wet weather favours the disease, with symptoms developing between four and eight days after infection when temperatures are optimal (18–20°C).

Importance
Although rarely seen in Queensland lettuce production areas, anthracnose may be an important disease in southern states during periods of cold, wet weather.

Management
- Destroy or incorporate crop debris by deep ploughing.

- Rotate lettuce with other crops and avoid planting crops in fields with a history of the disease.

- Time overhead irrigation programs to minimise periods of leaf wetness, which favour infection.

Fig 15.7 Severe symptoms of anthracnose showing the 'shot-hole' effect on a lettuce leaf.

- Apply the recommended fungicides. Base spray programs on regular applications of protectant fungicides. Only apply systemic sprays after periods of cool, wet weather when conditions are conducive to infection.

- Destroy alternative hosts.

■ BASE ROT

Cause
The fungus *Rhizoctonia solani*.

Refer to the chapter on Common diseases of vegetable crops.

■ BLACK ROOT ROT

Cause
The fungus *Thielaviopsis basicola*.

Symptoms
Plants are affected in both seedling nurseries and production fields. Infected plants are slow growing and remain stunted in comparison with healthy plants. Apart from the lack of vigour and general chlorosis, there usually are no specific above-ground symptoms.

Affected areas on roots are dark brown to black. Root lesions may be small or may coalesce to affect the whole root.

Root system volume in severely affected plants is greatly reduced, and, in extreme cases, roots are reduced to stubs.

Source of infection and spread
The fungus can affect a wide range of crops and is capable of surviving in soil for long periods. Common sow thistle (*Sonchus oleraceus*) is a weed host.

Fig 15.8 Black root rot (left) causing stunting and blackened roots.

Fig 15.9 Roots affected by black root rot are small and coloured brown to black.

Poor drainage and moderate soil temperatures (17–25°C) favour the disease. In Queensland, crops can be affected for most of the year, but high soil temperatures in summer reduce disease severity.

The area affected by the disease and disease severity increase gradually with successive crops.

Importance
Black root rot was first identified in Queensland lettuce crops in 1983. It is currently known to occur in the Brisbane metropolitan area, the eastern Darling Downs and the Granite Belt. The disease has also been recorded in New South Wales. Black root rot has the potential to reduce greatly the proportion of marketable heads from a crop.

Management
- Differences in varietal susceptibility occur with some varieties having good resistance.

- Avoid planting consecutive lettuce crops.

- Avoid close cropping with other susceptible crops such as soybean, cowpea, clover or lucerne.

- Ensure soil is free-draining.

- Pre-plant soil fumigation will give short-term control.

■ DOWNY MILDEW

Cause
The oomycete *Bremia lactucae*.

Fig 15.10 Downy mildew symptoms on upper and lower leaves of lettuce.

Fig 15.12 Downy mildew lesions on upper and lower (inset) leaf surfaces of a lettuce.

Symptoms

Light green to yellowish, angular spots develop on the upper surface of leaves, later turning brown and sometimes becoming soft and slimy. In moist weather, a white, downy, fungal growth is present on the underside of the spots.

Source of infection and spread

As lettuce is grown throughout the year, the main source of infection for young crops comes from nearby older lettuce crops, although the fungus can survive as oospores in soil. The downy growth produced on leaves in moist weather contains large numbers of spores that develop at night, and released in the early morning and dispersed by wind over long distances. Spores are short-lived and infection can occur in three to four hours if free moisture is present and temperatures are in the range of 10–22°C. Cool, moist weather favours disease development. Downy mildew is generally more severe on lower leaves, which are shaded, and remain wet for a considerable time after rain, irrigation or heavy dews.

Importance

Downy mildew is a common disease requiring specific control methods. Severely affected plants may require excessive trimming during packing, and are predisposed to secondary rot organisms causing losses during storage and transport. *Bremia lactucae* exists as a complex of races or strains that have the capacity to overcome host resistance genes or develop resistance to fungicides used in disease management.

Management

• Spray with the recommended protectant and eradicant fungicides. Base the he spray program on regular protectant fungicides (applied every seven to 10 days), with systemic products applied only following cool, wet periods.

Fig 15.11 Downy mildew symptoms on cos lettuce.

Fig 15.13 Downy mildew spore production on lettuce leaf.

FUSARIUM WILT – BIOSECURITY THREAT

Cause
The fungus *Fusarium oxysporum* f.sp. *lactucum*.

Symptoms
Infected seedlings wilt and die. In older plants, the oldest leaves turn yellow, wilt and then become necrotic. Often stunted, the plants may fail to form heads. Internally, there is a red-brown discolouration of the vascular tissues in the stems, in the cortex of the crown and in the taproot. Symptoms of this disease may be confused with those of ammonium toxicity or Verticillium wilt.

Source of infection and spread
Fusarium oxysporum f.sp. *lactucum* is a soil-borne fungal disease that can persist in infected soils for long periods due to the production of resilient chlamydospores. The disease tends to be most severe on lettuce planted during warmer production months. This pathogen is carried in seed and its introduction into Italy in 2002 was most likely by seed-borne transmission.

Importance
This disease has been reported in Asia (Taiwan, Japan, Korea), Europe (Italy) and the USA. It is a serious pathogen and a major threat to lettuce production in the areas in which it occurs. The potential for widespread transmission of this disease is significant due to its ability to survive in lettuce seed.

What to do if you suspect Fusarium wilt
This pathogen is a biosecurity risk to Australia. Any suspected affected plants should be reported to the nearest Department of Primary Industries or the Plant Health Australia hotline (1800 084 881).

Fig 15.14 Fusarium wilt in lettuce.

- Plough in old crops immediately after harvest.
- Resistant varieties are available.

■ GREY MOULD

Cause
The fungus *Botrytis cinerea*.

Symptoms
A water-soaked brown-grey soft rot develops on the lower leaves and stems or on damaged leaves. Tissues that are injured and wet or in contact with the soil are particularly susceptible. Grey fuzzy fungal growth, and in some cases, dark sclerotia, may be visible on the rotten tissues. Severely affected plant stems become girdled causing wilting and plant collapse.

Source of infection and spread
The fungus survives as a saprophyte on crop debris and weed hosts, as a pathogen on numerous hosts and as sclerotia in the soil. The fungus produces wind-borne spores that germinate and invade susceptible tissues if free moisture is available. Cool temperatures (18–23°C), and

Fig 15.15 Grey mould on lettuce.

Fig 15.17 Sclerotinia rot (lettuce drop).

Symptoms

A soft, watery rot develops, generally beginning at ground level and progressing into the head, causing the leaves to rapidly collapse and die. The head eventually becomes a wet mass of light brown, decayed leaves. Affected areas show masses of white, fluffy, fungal growth in which black sclerotia occur. The sclerotia of *S. sclerotiorum* are irregularly shaped and much larger (5–1 mm) than the more circular-shaped (3–5 mm) sclerotia of *S. minor*.

Source of infection and spread

Refer to the chapter on Common diseases of vegetable crops.

Fig 15.16 Grey mould on lettuce in the field.

foggy or showery conditions favour development. Lettuce tissues are predisposed to infection by frost or heat damage, physiological stress or attack by other pathogens.

Importance

Grey mould is generally a minor disease, but it may develop rapidly under cool, wet conditions.

Management

- Spray with the recommended protectant and systemic fungicides. Base the spray program on regular protectant fungicides (applied every seven to 10 days), and apply systemic products after cool, wet periods.

- Plough in old crops immediately after harvest.

■ SCLEROTINIA ROT (DROP)

Cause

The fungi *Sclerotinia sclerotiorum* and *S. minor*.

Importance

Sclerotinia rot is a common and often destructive disease of lettuce, particularly during cool, moist weather.

Management

- Avoid planting infested land in the cooler months.

Fig 15.18 *Sclerotinia minor* on lettuce.

Fig 15.19 Lettuce plant infected with *Sclerotinia* in the field.

Fig 15.20 Advanced Septoria leaf spot on lettuce.

- Rotate lettuce with non-susceptible crops, for example cereals for several years to reduce soil inoculum levels.

- Grow green manure crops that provide a biofumigant effect to reduce soil inoculum levels before re-planting lettuce.

- Avoid wet, shady areas.

- Grow varieties with an upright growth habit in high-risk situations.

- Spray with the recommended protectant and systemic fungicides. Base the spray program on regular protectant fungicides (applied every seven to 10 days), and apply systemic products following cool, wet periods. Where *Sclerotinia minor* is involved, apply a drench spray around the base of transplants. When *Sclerotinia sclerotiorum* is present, apply fungicides to the foliage as indicated above.

- Dispose of affected crop residues by deep cultivation immediately after harvest.

- Avoid irrigation close to harvest.

◼ SEPTORIA SPOT

Cause
The fungus *Septoria lactucae*.

Symptoms
Light brown, irregular spots develop predominantly on the lower leaves, and are studded with small, black fruiting bodies (pycnidia) of the fungus. As the disease progresses, the lesions desiccate and the dead leaf tissue in the centres of the spots may drop out, leaving holes or ragged edges.

Source of infection and spread
The fungus survives on crop residues and weeds, and introduced on infected lettuce seed. Spores produced in fruiting bodies spread during wet, windy weather.

Fig 15.21 Mild symptoms of Septoria leaf spot.

Importance
Septoria spot may be serious during prolonged periods of wet weather.

Management
- Plough in old crops immediately after harvest and do not re-plant lettuce in areas containing undecomposed crop residues.

- Spray regularly at seven to 10 day intervals with the recommended protectant fungicides.

- Rotate lettuce with other crops.

VIRUSES

◼ CUCUMBER MOSAIC

Cause
Cucumber mosaic virus (CMV; Cucumovirus).

Symptoms

Affected plants are stunted and leaves develop a yellow mottle, distortion and necrotic or dead areas.

Symptoms are similar to several other viruses infecting lettuce, for example lettuce mosaic, lettuce necrotic yellows and tomato spotted wilt viruses. Laboratory tests are often necessary to determine the virus(es) involved.

Source of infection and spread

CMV has a very wide host range among weed, crop and ornamental species. These hosts provide the inoculum, transmitted by aphids into lettuce. Transmission is non-persistent, so aphids acquire and transmit the virus during very short feeding times.

Importance

CMV is usually a minor problem of lettuce in Australia.

Management

- Control weed hosts adjacent to lettuce crops.

- Destroy old crops as soon as harvesting is completed.

■ LETTUCE BIG-VEIN DISEASE

Cause

Fungal-transmitted viruses e.g. *Mirafiori lettuce virus* (Ophiovirus).

Symptoms

Leaves on affected plants have enlarged, transparent veins. Plants are often stunted, with small hearts and upright, ruffled leaves. Young plants are most susceptible to infection.

Source of infection and spread

The virus spreads in a soil-borne chytrid fungus *Olpidium virulentus*. Swimming spores (zoospores) of the fungus infect lettuce roots and transmit the virus. As water is essential for zoospore movement, big-vein disease is common in crops grown on soils that retain moisture and in poorly drained areas. Lettuce big-vein is also common in hydroponic systems where the fungus contaminates nutrient solutions. The disease is favoured by cool temperatures and little symptom expression occurs above 22°C. Resting spores of the fungus containing the virus can survive in soil for over 20 years. Infected spores spread in drainage water, contaminated soil, infected transplants and contaminated crates and machinery.

The virus has a narrow host range including lettuce, endive and *Sonchus* (sowthistle) species.

Importance

Big-vein is a common and, at times, serious disease during cool weather particularly on heavier soils that remain wet for long periods. The disease may also be common in seedling production nurseries and hydroponic systems. Early infection can cause a considerable loss of yield and quality.

Management

- Produce and use healthy transplants.

- Treat potting media or seedlings with recommended fungicides to kill zoospores.

- Maintain a high standard of hygiene in nurseries and hydroponic systems.

- Avoid contaminating land with the fungus by attending to farm hygiene.

Fig 15.22 Big-vein disease on lettuce.

Fig 15.23 Big-vein disease in cos lettuce.

- Avoid poorly drained, heavy soils for lettuce production.

- Manipulate planting dates to avoid cool, overcast weather in locations where big-vein disease is likely to be serious.

- Reduce the incidence of disease and crop losses by using resistant varieties.

■ LETTUCE NECROTIC YELLOWS

Cause
Lettuce necrotic yellows virus (Rhabdovirus).

Symptoms
Initial symptoms are a browning of leaf veins, followed by partial death of the inner leaves. Affected plants are yellow and stunted, often with twisted and lopsided leaves. In advanced stages, the outer leaves wilt severely, giving the plant a flattened, stunted appearance.

Source of infection and spread
The weed common sowthistle (*Sonchus oleraceus*) is the major host of both the virus and the sowthistle aphid (*Hyperomyzus 1actucae*), which spreads the virus in a persistent and propagative manner. This aphid is the major carrier (vector) of the virus and often found in large numbers on the flower stalks of sowthistle.

Outbreaks of necrotic yellows are usually associated with infected sowthistles within or near lettuce crops. The sowthistle aphid does not breed on lettuce, and there is little plant-to-plant spread of the virus in lettuce crops.

Importance
Necrotic yellows is a common disease which can cause serious sporadic losses.

Management
- Destroy sowthistle weeds in and around the crop.

- Maintain aphid control in crops.

■ MOSAIC

Cause
Lettuce mosaic virus (Potyvirus).

Symptoms
Light green and dark green mosaic patterns develop on young leaves, sometimes with vein browning. Leaf margins may have a ragged or frilly appearance. Affected plants are stunted, pale in colour, and generally do not produce marketable heads.

Source of infection and spread
The virus is both seed-borne and spread from plant to plant by aphids in a non-persistent manner. Seed-borne infection can be up to 3% in badly contaminated seed lots, with levels as low as 0.1% being sufficient to initiate a serious epidemic.

The virus is spread by aphids, with only short feeding periods (less than one minute) being necessary for transmission. The virus can spread by many species of aphids occurring in or around a lettuce field. Considerable spread may occur from aphids migrating through a crop, making only brief probes on plants as they move. Planting overlapping lettuce crops and failing to destroy old plantings with high disease levels are major factors in the spread of the virus. Insecticides are of little value for mosaic control because of the very short feeding periods required by aphids for virus transmission.

Fig 15.24 *Lettuce necrotic yellows virus.*

Fig 15.25 *Lettuce mosaic virus* on lettuce leaves.

Fig 15.26 *Lettuce mosaic virus* in a field plant.

Apart from lettuce, *lettuce mosaic virus* has a range of weed hosts that may provide a source of infection for lettuce crops.

Importance
Lettuce mosaic is relatively common and can cause serious losses when high disease levels occur.

Management
- Use tested seed that is free from *Lettuce mosaic virus*.

- Destroy old lettuce crops as soon as harvesting is completed.

- Avoid planting new crops adjacent to old crops.

■ SPOTTED WILT

Cause
Tomato spotted wilt virus (Tospovirus).

Symptoms
Affected plants are yellow and stunted with marginal wilting and a lopsided appearance. Brown spots and streaks develop on leaves and petioles. The younger internal leaves and petioles may develop small, light brown necrotic spots or turn dark and rot from secondary infection by soft-rotting bacteria.

Plants infected when young may collapse and die.

Source of infection and spread
Refer to the chapter on Common diseases of vegetable crops.

Fig 15.27 *Tomato spotted wilt virus* in lettuce.

Importance
Tomato spotted wilt virus is a major cause of crop failure and loss in the Australian lettuce industry.

Management
- Refer to the chapter on Common diseases of vegetable crops.

■ TURNIP MOSAIC

Cause
Turnip mosaic virus (Potyvirus; TuMV).

Symptoms
Early symptoms are small, light green, circular to irregular spots on leaves. The midrib often curves and the leaves become distorted and chlorotic. Affected plants are stunted.

Fig 15.28 *Turnip mosaic virus* on lettuce.

Source of infection and spread

Aphids spread the virus in a non-persistent manner. TuMV has a wide host range among crop and weed species. It frequently infects cultivated and weedy brassicas in cooler regions.

Importance

The virus is not usually an important disease in lettuce crops. Susceptibility to TuMV is restricted to some crisphead varieties and is linked to a downy mildew resistance gene.

Management

• Avoid planting lettuce adjacent to brassica crops in areas where TuMV can be a problem.

• Avoid lettuce varieties highly susceptible to TuMV in these areas.

■ FURTHER INFORMATION

Davis RM, Subbarao KV, Raid RN and Kurtz EA (Eds) (1997). *Compendium of lettuce diseases*. APS Press, St Paul, Minnesota.

Koike ST, Gladders P and Paulus AO (2007). 'Lettuce', in *Vegetable diseases – a colour handbook,* pp. 296–326. Academic Press, New York.

Latham LJ, Jones RAC and McKirdy SJ (2004). Lettuce big-vein disease: Sources, patterns of spread, and losses. *Australian Journal of Agricultural Research* **55**, 125–130.

Latham LJ and Jones RAC (2004). Deploying partially resistant genotypes and plastic mulch on the soil surface to suppress spread of lettuce big-vein disease in lettuce. *Australian Journal of Agricultural Research* **55**, 131–138.

Onion and related crops are monocotyledons; they are members of the genus *Allium* and belong to the Alliaceae family and are used widely for cooking and seasoning. The most important *Allium* crop in Australia is onion (*Allium cepa*); smaller but significant areas of garlic (*A. sativum*), leek (*A. porrum*), chives (*A. schoenoprasum*) and shallot (*A. cepa* var. *ascalonicum*) are grown. All States grow onions and South Australia and Tasmania are the largest producers. Important diseases include downy mildew, neck rot and the soil-borne disease, white rot.

BACTERIA

■ BACTERIAL BLIGHT

Cause
The bacterium *Pseudomonas syringae* pv. *porri*.

Symptoms
The disease is usually only a problem in leek.

Young leaves become water-soaked and develop yellow, longitudinal lesions or stripes. The leaves can become twisted and curled as growth continues.

Older leaves develop brown leaf lesions surrounded by a narrow, yellow halo.

Seedling plants can be severely affected resulting in death and poor establishment.

Fig 16.2 Bacterial blight leaf twisting symptoms.

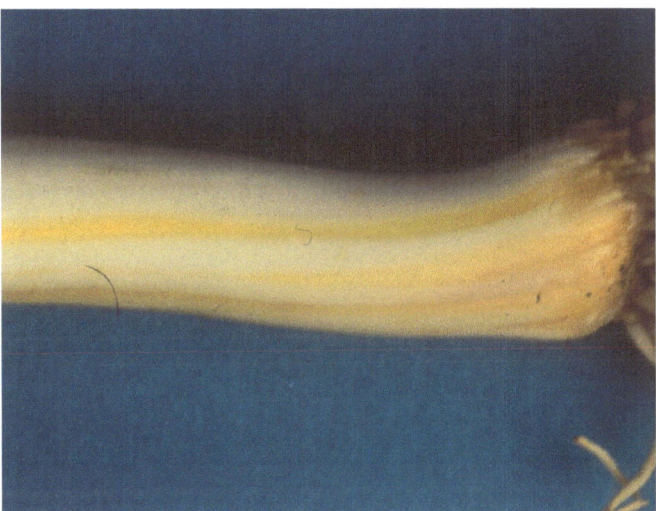
Fig 16.1 Bacterial blight in onion leek.

Fig 16.3 Leaf lesions caused by bacterial blight.

Mature leeks may have yellow to pale brown streaks running the length of the bulb.

Source of infection and spread

The bacterium is seed-borne and often introduced into direct-seeded crops or on leek transplants. The bacterium spreads in the nursery or field is in water splash from rain or overhead irrigation. The bacterium can survive on undecomposed crop residues.

Importance

Bacterial blight is a common disease of leek in Australia. The disease causes losses in seedling plants, and streaking on mature plants makes them unmarketable.

Management

• Avoid over-head irrigation to reduce or eliminate the pathogen.

• Do not plant seed potentially contaminated with the pathogen.

• Regular, protectant copper sprays used to control other foliar diseases will also help to limit the spread of this disease.

XANTHOMONAS LEAF BLIGHT – BIOSECURITY THREAT

Cause

The bacterium *Xanthomonas axonopodis* pv. *allii*.

Symptoms

Initially, symptoms on leaves appear as irregularly shaped white flecks and spots, sometimes surrounded by water-soaked margins. The lesions enlarge rapidly; become tan to brown and water-soaked. Chlorotic streaks, often extending the length of the leaf, develop on some varieties. As the disease progresses, lesions coalesce causing tip dieback and extensive blighting of older, outer leaves. Under favourable conditions, all the leaves may be completely blighted, and plants senesce early.

Source of infection and spread

The bacterium can disperse on onion seed and bulbs. It survives in crop debris, contaminated seeds and as either a pathogen or an epiphyte on *Allium* species, weeds and legumes, particularly bean. The pathogen spreads in water splash, wind driven rain, surface water movement and on contaminated farm equipment and clothes.

Warm or hot weather and humid, wet weather favour the disease.

The bacterium infects plants through natural openings such as stomata, lenticels and wounds.

Fig 16.4 Xanthomonas leaf blight causing tip death and water soaking in the leaves.

Fig 16.5 Advanced symptoms of Xanthomonas leaf blight.

Importance

Although onion is most affected, the disease also occurs on chive, garlic, leek, shallot and Welsh onion. The disease was first reported from Hawaii in 1978 and has spread rapidly to many onion-producing regions in Asia, North America, South America, Japan and South Africa. Yields losses depend on the time of infection, weather conditions and the variety. Yield reductions of 20% are common and seedling infection in tropical climates can result in crop failure.

What to do if you suspect Xanthomonas leaf blight

This pathogen is a biosecurity risk to Australia. Any suspected affected plants should be reported to the nearest Department of Primary Industries or the Plant Health Australia hotline (1800 084 881).

FUNGI

■ BLACK MOULD

Cause

The fungus *Aspergillus niger*.

Symptoms

Black, powdery fungal growth develops on and between the bulb scales. In severe cases, the entire surface of the bulb may turn black and bulb rot develops.

Fig 16.6 Black mould on the outer scale leaves of an onion bulb.

Source of infection and spread

The fungus is common in soil. Spores are present on onion seed and the fungus survives as a saprophyte on crop residues and decayed plant material. Spores disperse in wind and infect senescing onion leaves and then bulb tissues. Bulb infection in the field is usually associated with injury. The fungus requires high humidity and free water for six to 12 hours on the onion surface for infection to occur.

Importance

The disease is most likely to occur during warm to hot, humid weather and can cause serious losses during transit and storage.

Management

• Rotate onions with crops that do not belong to the Alliaceae.

• Harvest bulbs during dry weather and minimise damage to the neck tissue of bulbs.

• Store the bulbs in cool, dry conditions.

BOTRYTIS LEAF BLIGHT – BIOSECURITY THREAT

Cause
The fungus *Botrytis squamosa* (telemorph *Botryotinia squamosa*).

Symptoms
The first symptoms are small leafspots with a white centre surrounded by a light green halo about one millimetre wide. The presence of the halo is diagnostic for the disease and allows it to be distinguished from damage and insect feeding punctures. The spots may expand, become elliptical in shape and the halo may disappear. The disease develops rapidly under prolonged moist conditions and moderate temperatures, causing bleached foliage and leaf collapse. Severely affected crops have a blighted appearance due to leaf desiccation and subsequent death.

Source of infection and spread
The fungus survives as sclerotia in the soil, in crop debris and on diseased crops. The hosts of *Botrytis squamosa* are restricted to *Allium* species. The fungus disperses in air-borne spores (conidia) and, for infection to occur, at least six hours of leaf wetness are required and temperatures in the range 12–24°C. The extent of disease development and leaf blighting depends on the number of hours of leaf wetness. The longer the leaves remain wet the more severe the leaf blighting. Dense crops favour disease development.

Importance
Botrytis leaf blight occurs in most countries where *Allium* species are grown and is a destructive disease during prolonged periods of wet weather. The pathogen has not been found in Australia.

What to do if you suspect Botrytis leaf blight
This pathogen is a biosecurity risk to Australia. Any suspected affected plants should be reported to the nearest Department of Primary Industries or the Plant Health Australia hotline (1800 084 881).

Fig 16.7 Botrytis leaf blight.

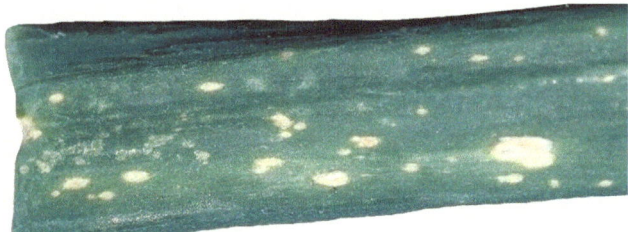

Fig 16.8 Lesions of Botrytis leaf blight.

■ DOWNY MILDEW

Cause
The oomycete *Peronospora destructor*.

Symptoms
The disease often appears in localised areas in a field and then spreads to surrounding areas.

Early symptoms include leaf tip bleaching and small, irregularly shaped blotches on the leaves. Lens shaped spots develop on leaves and become covered by a blue-grey fungal growth during humid periods. Under severe conditions, whole leaves are affected. The leaves gradually become pale green and later yellow. Diseased parts of leaves, especially the tips, may collapse. Downy mildew is often more severe in patches that were the sites of early

Fig 16.9 Downy mildew in the field.

infections. Secondary pathogens such as *Stemphylium* and *Alternaria* commonly infect the affected leaf tissue.

Source of infection and spread

The pathogen survives between seasons as thick-walled spores (oospores) in onion crop residues and as oospores and mycelium inside bulbs. During the growing season, large numbers of spores (conidia) produced by the pathogen at night on infected leaves, are spread by wind currents during the day. Cool weather with light winds favour spore dispersal and spores can survive on leaf surfaces for up to three days. The pathogen requires cool temperatures (less than 22°C) and free moisture on the leaves for at least three hours before infection can occur. Severe disease outbreaks are most likely to occur during periods of cloudy, mild days and cool, still, dewy nights. Dry, sunny weather usually halts disease progress.

Importance

Downy mildew is a destructive disease of onions, leeks, garlic and related species during periods of cool, humid weather.

Management

- Incorporate crop residues after harvest.

- Rotate onions and related species with other vegetable or field crops.

Fig 16.10 Blue-grey growth of downy mildew on an onion leaf.

- Apply recommended protectant and eradicant fungicides, particularly before and during periods that favour downy mildew development. Use disease-forecasting systems where available to enhance fungicide application efficiency.

- High-density plantings provide good conditions for disease build-up and make spray coverage more difficult. Wide row spacing allows air movement through the crop canopy. Balance the yield advantages of high densities with the possibility of increased downy mildew severity.

■ FUSARIUM WILT, BASAL PLATE AND CROWN ROT

Cause

Several species of *Fusarium* including *F. oxysporum*, *F. avenaceum* and *F. culmorum*.

Symptoms

Leaf symptoms can develop at any stage during growth. A yellowing and death of leaves occurs, generally starting with the younger leaves. A wet breakdown at the neck may also develop and the bulb may become distorted or bloated. With all affected species, a basal plate rot occurs where the roots attach to the crown. The affected area is water-soaked, light tan to brown and initially the tissues are firm, although a soft rot may develop as the disease progresses. The roots of affected plants are poorly developed. Additional symptoms on onions include a tan to pink root rot; a red-purple discolouration may appear on the stems and bulbs of garlic; and on leeks, a brown-yellow or pink rot may be visible at the base of the stem or on the basal plate.

Source of infection and spread

Species of *Fusarium* are common soil inhabitants and persist in soils as resistant chlamydospores. The fungi may be introduced into new areas on diseased bulbs and garlic

Fig 16.11 Field symptoms of Fusarium wilt.

Fig 16.12 Onion plants infected with Fusarium wilt.

Fig 16.13 Onions infected with Fusarium wilt (left) compared with a healthy plant. Symptoms include wet breakdown in the neck and leaf.

cloves used as planting material and in infested soil adhering to farm machinery.

The fungi invade the roots and spread to the basal plate. Direct infection of bulbs may also occur. Disease development is favoured by temperatures of 25–28°C. Damage from insect feeding and wet weather near harvest can favour disease development.

Importance
The diseases caused by *Fusarium* are important in most *Allium* crops, including onion, leek, garlic and shallot.

Management
- Rotate onion and related crops with non-susceptible crops using a four-year rotation.

- Use resistant or tolerant varieties.

- Apply recommended fungicides as seedling drenches or as furrow treatments.

- Ensure seedlings are disease free.

Fig 16.14 Brown discolouration of the basal plate of an onion bulb infected with Fusarium wilt (left).

■ MALLEE ONION STUNT

Cause
The cause of this disease is still under investigation, but *Rhizoctonia solani* is the main pathogen.

Fig 16.15 Patchy growth in a crop affected by mallee onion stunt.

Symptoms

Affected onion crops develop areas of stunted plants. As crops mature, plants remain stunted and fields develop a patchy appearance. Reduced yields and bulb sizes occur in affected crops.

Source of infection and spread

Although several soil-borne fungi have been isolated from affected plants, there is a consistent association between the disease and the presence of *Rhizoctonia solani* anastomosis group 8. This fungus also causes a similar bare patch disease in cereals.

Importance

Mallee onion stunt causes economic losses in crops grown on the sandy floodplain soils along the Murray River in South Australia, particularly when cereal crops have been grown in rotation with onions.

Fig 16.16 Variation in plant growth caused by mallee onion stunt disease.

■ NECK AND BULB ROT

Cause

The fungi *Botrytis allii*.

Symptoms

Neck rot is primarily a disease of onion bulbs in storage. A soft breakdown of the cut neck and other injured parts of the bulb occurs, accompanied by the production of a dense layer of grey, furry mould. The fungal growth only becomes apparent after several weeks in storage. Hard, black resting bodies (sclerotia) of the fungus may become matted around the rotted neck. The rot continues to spread from the neck into the bulb and the entire bulb eventually becomes soft and rotted.

Source of infection and spread

The fungus survives in the soil as sclerotia, on crop residues or on onion seed. With seed-borne inoculum the pathogen spreads from the seed coat to seedling leaves while the seed coat remains attached to the cotyledons. Sclerotia in the soil germinate to produce air-borne conidia. Field symptoms

Fig 16.17 Onion neck and bulb rot.

Fig 16.18 Neck rot caused by the fungus *Botrytis*.

seldom develop. Infection occurs at harvest through the moist cut neck but may also develop through other injuries to the bulb. Botrytis can also grow downwards from the tips of leaves into leaf sheaths that form the storage scales at the neck of bulbs. Breakdown then occurs during storage, although this may be quite slow. Under humid storage conditions, extensive spread of the infection to other bulbs may occur by contact.

Importance

Neck and bulb rot can cause considerable losses in stored onions when storage conditions favour the fungus.

Management

- Use onion seed free of the pathogen or treat seed with recommended fungicides.

- Harvest at the correct stage of maturity.

- Maintain a relatively long neck on onions at harvest and thoroughly dry the bulbs before storage.

- Handle bulbs carefully to reduce injury.

- Store bulbs in cool, dry conditions.

■ PINK ROOT

Cause

The fungus *Phoma terrestris* (syn. *Pyrenocheata terrestris*).

Symptoms

Affected plants are unthrifty with yellowing, withering and death of leaves, starting from the tips of older leaves.

The term 'pink root' aptly describes the characteristic root symptom, as roots turn pink then darken in colour to red or purple and finally brown or black. Root death occurs, appearing first near the centre of the basal plate. Secondary organisms may enter through damaged tissues and cause bulb rot.

The symptoms of pink root may be confused with those caused by Fusarium basal rot.

Source of infection and spread

The fungus is a common soil inhabitant and most active in warm weather; soil temperatures of 24–28°C are optimum

Fig 16.19 Pink root in onion.

for disease development. There is little disease development in cool soils. Onion root tips are infected by direct penetration of hyphae and the fungus grows throughout the roots but does not invade the basal plate or fleshy scales of the bulb. The fungus has a wide host range including cereals, cucurbits, tomato and capsicum and it is able to survive in soil as chlamydospores and pycnidia, or in crop residues.

Importance

Pink root is a serious disease of onions grown in tropical and subtropical regions.

Management

- Avoid warm-weather plantings where the disease is known to occur.

- Rotate onions with non-susceptible crops. The disease is most severe where onions are grown continuously or in a one-year rotation.

- Fumigate or use solarisation on soils to reduce inoculum levels.

■ PURPLE BLOTCH

Cause

The fungus *Alternaria porri*.

Symptoms

Symptoms begin as small, water-soaked leaf spots that quickly develop white centres. The spots gradually enlarge, become zonate and brown to purple in colour. Older spots may extend around the leaves causing collapse and death. Bulbs can be infected when 'topped' at harvest, causing storage rots.

Source of infection and spread

The fungus survives in crop residues. Spores are air-borne and infect leaves via the stomata or directly through the epidermis. Warm, humid or wet weather favours disease development. Leaves need to be wet for eight hours at temperatures of 15–25°C for infection to occur. Older leaves are more susceptible to infection than younger leaves and damage from thrips feeding increases the susceptibility of both young and older leaves to the fungus.

Importance

Purple blotch is a common leaf disease of onions, garlic and leek grown in warm, humid areas.

Fig 16.20 Purple blotch symptoms in onion.

Management

- Incorporate crop residues after harvest to reduce inoculum levels.

- Rotate crops with species outside of the Alliaceae family.

- Minimise the hours leaves are wet if using overhead irrigation.

- Apply regular applications (every seven to 10 days) of the registered, protectant fungicides. This is particularly important as the crop canopy ages and becomes dense, and when weather conditions favour infection.

Fig 16.21 Purple blotch symptoms on onion leaves.

Fig 16.22 A large purple blotch lesion.

Fig 16.23 Rust in garlic.

■ RUST

Cause
The fungus *Puccinia allii*.

Symptoms
Small, white flecks and spots develop on leaves and stems. As the disease progresses, the flecks develop into oblong lesions, 1–3 mm in length, containing masses of orange spores that erupt through the epidermis forming pustules. Dark brown teliospores may form later in the season. Severely affected leaves turn yellow, wilt and die.

Source of infection and spread
Fungal spores survive on crop residues, volunteer plants and older infected plantings of onions and related species. Spores can disperse over considerable distances in the wind. Disease development depends on several factors including available inoculum, temperature and relative humidity. Temperatures of 12–21°C and high relative humidity for at least four hours are optimum conditions

for infection. Dense plantings, high nitrogen applications and low potassium levels favour disease development.

Importance
Rust occurs throughout the world on most *Allium* species and can cause considerable damage in onion, leek, garlic and chive crops. *Puccinia allii* is considered to be a complex of related species that vary in host range and several morphological characters. For example, rust collections from leek and garlic generally have distinct host ranges among cultivated *Allium* species.

Management
- Destroy crops after harvest and incorporate crop residues into the soil.

- Rotate *Allium* crops with unrelated species.

- Separate new plantings from maturing crops.

Fig 16.24 Smudge on the outer scale leaves.

Fig 16.25 Stemphylium leaf blight in onion.

■ SMUDGE

Cause

The fungus *Colletotrichum circinans*.

Symptoms

The disease mainly affects the dried wrapper scales and lower portions of white onion. Small, round, dark blotches develop on the bulb and frequently form circular, concentric rings, approximately one centimetre in diameter. Smudge may be confused with smut, which forms dark streaks rather than spots; and with white rot.

Source of infection and spread

The fungus survives in the soil on onion residues. Spores are carried on the surface of the bulb when harvested, and symptoms develop if bulbs are stored for long periods or under poor storage conditions. Warm, wet weather favours infection.

Importance

Smudge affects onion, shallot and leek. The disease seldom affects coloured varieties of onions.

Management

• Rotate onion and related crops with other crops.

• Use clean seeds or healthy transplants.

• Cure bulbs quickly and store in cool, dry conditions.

■ STEMPHYLIUM LEAF BLIGHT

Cause

The fungi *Stemphylium vesicarium* and *S. botryosum*.

Symptoms

White flecks develop on leaves and enlarge to produce purple sunken lesions often surrounded by a yellow to pale brown border. The symptoms are similar to those produced by *Alternaria porri*, the cause of purple blotch.

Source of infection and spread

The fungi survive in infested crop residues, and spores disperse in the wind. Warm, humid conditions favour infection and disease development. *Stemphylium botryosum* is also a common saprophyte, infecting tissue damaged by other diseases, frost, and mechanical abrasion.

Importance

Both these species of *Stemphylium* are found on several *Allium* species in Australia, including onion, garlic and leek. Damage can range from mild leaf symptoms to leaf blighting following prolonged periods of wet weather.

Management

• Destroy harvested crops and incorporate crop residues into the soil.

• Plant resistant or tolerant varieties in areas where the disease is expected to be a problem.

• Apply regular applications of mancozeb at seven to 10 day intervals, particularly when weather conditions favouring disease development are likely.

■ WHITE ROT

Cause

The fungus *Sclerotium cepivorum*.

Fig 16.26 An onion field affected by white rot.

Fig 16.28 Micro- and macro-sclerotes of white rot (*Sclerotium cepivorum*).

Symptoms

Plants die out in small patches that extend slowly, often advancing only half a metre in a season. When removed from the soil, white wefts of fungal growth are visible on the bulbs and roots of affected plants. This growth causes soil particles to adhere to the affected areas. Small, dark brown, spherical resting bodies (sclerotia) of the fungus are embedded in the tissues. If affected bulbs are packed, extensive breakdown may occur during transit and storage.

Source of infection and spread

Two sclerotial forms of the fungus are present in Australia. The most common of these produces dark spherical sclerotes 0.2–0.5 mm in diameter, and the other produces larger sclerotes 0.5–2.0 cm in diameter. The sclerotes of both forms enable the fungus to survive in the soil for at least 20 years in the absence of host plants. The sclerotes germinate in response to chemical exudates from the roots of *Allium* species and the fungal mycelium invades the roots and basal plate. Secondary spread can occur as

mycelium grows from plant to plant when roots are in close contact. Cool weather favours the disease and temperatures below 20°C are optimum for the fungus. Introduction of the disease to new areas may occur by the transport of sclerotia in infested soil or diseased seed bulbs.

Importance

White rot is a major and widespread disease of onion when grown intensively during the cooler months of the year. All *Allium* species can be affected, but onion and garlic are the most susceptible. The persistence of the pathogen in the soil makes management of white rot difficult.

Management

- Select fields free from the pathogen and take precautions to prevent introduction of the fungus. The fungus can be introduced on transplants, contaminated machinery and equipment and in drainage water.

- Do not plant onions or garlic in affected areas for at least 10 years.

- Do not retain bulbs from a diseased crop for planting.

- Apply the recommended fungicides at sowing and during crop growth.

- Fumigate affected areas to reduce inoculum levels.

VIRUSES

■ IRIS YELLOW SPOT

Cause

Iris yellow spot virus (IYSV; Tospovirus).

Fig 16.27 White rot symptoms in harvested plants.

Fig 16.29 *Iris yellow spot virus* causes lens-shaped lesions on onion leaves and seed stalks.

Symptoms

Symptoms include dry, straw coloured, spindle or diamond-shaped lesions on leaves or seed stalks (scapes). The lesions often develop on the margin of the youngest, fully developed leaves or on the swollen part of the scape. The lesion centres may have green centres and concentric rings of alternating green and straw-coloured tissue sometimes occur. As lesions expand and age, tissue death occurs and the lesions are often colonised by saprophytic fungi or scalded by sunlight, giving a red or feint purple colour. These symptoms may resemble those caused by downy mildew or purple blotch. Large areas of foliage may die and the large necrotic areas on the seed stalks cause lodging and loss of seeds.

Source of infection and spread

The natural hosts of the *Iris yellow spot virus* (IYSV) appear to be largely restricted to members of the Allicaceae family and include most *Allium* species, alstroemeria, hippeastrum, iris, lisianthus and pelagonium.

IYSV is maintained in onion seed and bulb crops, volunteer plants, infected transplants, weed hosts and infected ornamental species.

The virus is transmitted by onion thrips (*Thrips tabaci*) with the transmission process being very similar to that occurring with other tospoviruses (refer to the tospovirus section in the chapter on Common diseases of vegetable crops). First and early second stage thrips larvae acquire the virus, which circulates and replicates in the insect before being transmitted by larvae and adults. Other thrips species are not known to transmit the virus and transmission through seed has not been demonstrated.

Importance

IYSV, first described in 1998, has now been identified in many countries, including Australia. The virus can be very damaging in seed crops through lodging of the seed stalk and in onion production through a reduction in yields and quality of bulbs.

Management

- Destroy harvested crops and volunteer plants.

- Maintain uniform and dense plant populations as these have been shown to reduce the incidence IYSV.

- Plant varieties with tolerance or resistance to IYSV.

- Separate young crops from older crops.

- Manage thrips populations by insecticides and other means.

■ MOSAIC

Cause

Several viruses cause mosaic symptoms in onion, garlic, leek and related crops in Australia. These include *Onion yellow dwarf virus*, *Leek yellow stripe virus* and *Shallot latent virus*.

Symptoms

Infected plants develop yellow stripes and flecking, blotching and streaking of leaves. Affected plants are generally stunted.

Source of infection and spread

The viruses may be carried from season to season in diseased garlic cloves and in onion bulbs and sets. The viruses also survive in harvested crops and volunteer plants.

Fig 16.30 Mosaic symptoms in onion.

Aphids can transmit the viruses from plant to plant in a non-persistent manner.

Importance
Virus infection can occur in all cultivated *Allium* species in Australia, particularly garlic.

Management
- Maintain and plant healthy garlic bulbs.

- Destroy infected crops after harvest.

- Separate new plantings from older plantings.

■ FURTHER INFORMATION

Brewster J (2008). *Onions and other vegetable Alliums*. CABI Publishing, Wallingford.

Hall BH, Hitch CJ, Oxspring EA and Wicks TJ (2007). Leek diseases in Australia. *Australasian Plant Pathology* **36**, 383–388.

Schwartz HF, Krishna Mohan S (Eds) (2008). *Compendium of onion and garlic diseases and pests* (2nd edn). APS Press, St Paul, Minnesota.

Parsnip (*Pastinaca sativa*) is a hardy, winter-grown root vegetable belonging to the Apiaceae family, which also includes carrot, celery, parsley and many herbs. Parsnip produces a long, tapering, storage taproot, usually with a smooth surface. The crop prefers a deep, friable, well-drained soil. The crop matures in six to nine months, depending on the temperatures during growth.

BACTERIA

■ CROWN GALL

Cause
The bacterium *Agrobacterium tumefaciens*.

Symptoms
Brown overgrowths develop on the crown and roots near the soil line. Affected plants may become stunted, produce small, chlorotic leaves and are more susceptible to adverse environmental conditions, especially cold or frost.

Source of infection and spread
The bacterium has a wide host range, encompassing at least 140 genera across 60 plant families. It overwinters in infested soils, where it can live as a saprophyte for several years. Parsnips grown in infested soils become infected when the bacterium enters the roots or crowns through wounds made by mechanical damage, insects and/or growth splits. The bacterium causes the host cells to divide quickly. This rapid, undifferentiated cell division produces the galls.

Importance
Crown gall has a worldwide distribution on many crops, but it is only a minor disease on parsnip in Australia.

Management
- Control measures for crown gall are generally not required in parsnip crops.

FUNGI

■ ITERSONILIA CANKER (BLACK CANKER)

Cause
A complex of pathogens, primarily: *Itersonilia perplexans*, *Fusarium* spp. and *Cylindrocarpon* spp. Other pathogens, including: *Mycocentrospora acerina*, *Acremonium* sp., *Microdochium* sp., *Phoma* sp., *Pythium sulcatum* and *Rhizoctonia* sp. can also cause cankers.

Symptoms
Red-brown to black cankers may develop on any part of the parsnip root. Most often, symptoms predominate in the

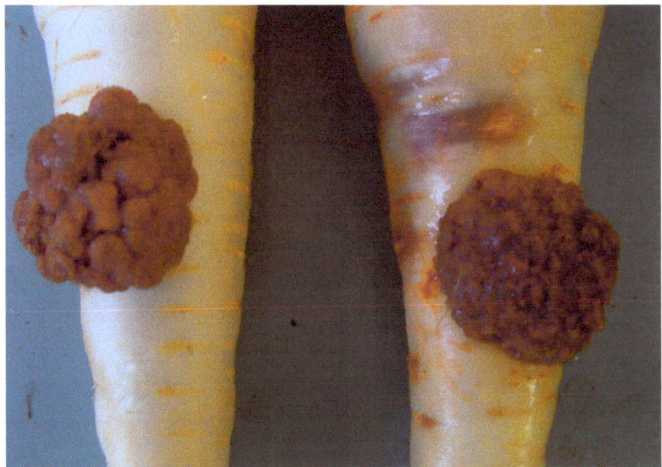
Fig 17.1 Symptoms of crown gall on parsnip roots.

Fig 17.2 Foliar symptoms of Itersonilia canker (black canker) in parsnip.

Fig 17.3 Black canker symptoms on parsnip crowns and roots.

crowns and/or shoulders of the roots. Root cankers are initially superficial but may affect deeper tissues with age. On leaves, small (1–2 mm diameter), irregular, brown spots with pale green to yellow haloes develop. Petioles, inflorescences and seed may also be affected.

Source of infection and spread

The main pathogens survive in the soil for many months as chlamydospores and in crop residues. Air-borne spores are also produced, released and dispersed, landing on leaves and petioles, giving rise to foliar infection. Spores wash from leaf lesions onto crown and upper root tissues leading to canker lesions. Cankers generally develop late in the growing season after extended rain periods. Infection by the *Itersonilia* fungus is favoured by cooler temperatures (optimum 20°C) and superficial damage to the roots predisposes parsnips to fungal attack and canker. Both *Phoma* and *Itersonilia* may be seed-borne.

Importance

Itersonilia canker is the major disease of parsnips in Australia, causing up to 80% crop loss in severe cases. The disease is most severe on spring-harvested crops (September to November).

Management

- Remove roots and plant trash from beds following harvest to minimise disease carry-over between crops.

- Practice a minimum rotation of 12 months between parsnip crops.

- Avoid planting parsnips in February or March since canker levels are historically highest during these plantings when harvested in the spring.

- Plant tolerant varieties if possible. The commercial cultivar 'Tusk' showed high emergence rates and low canker incidence in recent trials in Victoria.

■ POWDERY MILDEW

Cause

The fungus *Erysiphe heraclei*.

Symptoms

A superficial white fungal growth appears first on the older leaves and petioles. Younger leaves become infected as the disease progresses. When powdery mildew is severe, both upper and lower leaf surfaces are affected, the foliage becomes chlorotic and leaf senescence may occur prematurely.

Source of infection and spread

The fungus produces masses of air-borne spores that facilitate the spread of the pathogen between crops. Concurrent parsnip plantings, parsnip volunteers and weed hosts are potential sources of infection. *Erysiphe heraclei* attacks many members of the Apiaceae plant family, although host specialisation limits the number of hosts that particular isolates can infect. The disease cycle can complete within seven days under optimal conditions when high humidity and mild temperatures prevail. Disease severity increases under conditions of drought stress, and host susceptibility increases with age. Rain or overhead irrigation limits the disease.

Importance

Powdery mildew is a common disease of parsnips but it rarely causes yield losses. It is only significant cosmetic problem at harvest.

Management

- Destroy and remove old crops promptly after harvest.

- Apply the recommended fungicides.

Fig 17.4 Phloeospora leaf spot.

PHLOEOSPORA LEAF SPOT

Cause

The fungus *Phloeospora herclei*.

Symptoms

Small (1–2 mm diameter) pale green or brown leaf spots develop and coalesce to form large grey to brown necrotic patches. Extensive leaf death and defoliation may occur in severe cases. Root symptoms have not been reported.

Source of infection and spread

Little is known about the lifecycle of this pathogen. Fungal spores spread in splashing water and by direct leaf contact. The fungus presumably survives on infected crop residues and on alternate weed hosts.

Importance

Phloeospora leaf spot is a minor disease of parsnip in Australia.

LEAF BLIGHT

Cause

The fungus *Pseudocercosporella pastinacae*.

Symptoms

Angular to semicircular spots develop on the leaves. The spots are initially yellow but they progressively become brown and necrotic. Leaflets may die if foliar infection is severe. Lesions may also occur on petioles.

Source of infection and spread

There is no information available on the epidemiology of this pathogen.

Importance

Leaf blight is a very minor disease of parsnip in Australia.

DAMPING-OFF

Cause

The oomycete *Pythium* spp.

Refer to the chapter on Common diseases of vegetable crops.

RAMULARIA LEAF SPOT

Cause

The fungus *Ramularia pastinacae*.

Symptoms

Small (3–7 mm diameter), pale brown circular spots with darker margins and chlorotic haloes develop on the leaves. The centres of the lesions often fall out with age, giving the leaf a 'shot-hole' effect. White fungal sporulation may develop on the undersides of young lesions.

Source of infection and spread

Warm, wet conditions favour infection. The disease occurs in small foci and then when conditions favour infection it

Fig 17.5 Ramularia leaf spot.

Fig 17.6 Sclerotinia rot. Note the grey-white fungal mycelium and black 'seed-like' sclerote.

spreads to give a more general infection of the crop. Spores are most likely wind-borne.

Importance
Ramularia leaf spot is a common disease of parsnip, but it is rarely of economic importance.

Management
- Avoid sequential plantings of parsnip to minimise inoculum carry over to new parsnip plantings.

- Plough in old crops immediately after harvest and control parsnip volunteers.

- Regular foliar copper sprays should limit the development of this disease.

■ SCLEROTINIA ROT

Cause
The fungus *Sclerotinia sclerotiorum*.

Refer to the chapter on Common diseases of vegetable crops.

Pea (*Pisum sativum*) is a member of the legume family Fabaceae. Almost the entire Australian green pea crop is processed into frozen product. Snow peas are a popular fresh market crop. Bacterial blight and powdery mildew can be serious in pea crops while snow peas suffer severe losses from Fusarium wilt.

BACTERIA

■ BACTERIAL BLIGHT

Cause
The bacterium *Pseudomonas syringae* pv. *pisi*.

Symptoms
Bacterial blight can affect all above-ground parts of the plant. Lesions on leaflets and stipules are dark brown and irregularly shaped. They are water-soaked in the early stages, becoming papery as the tissue dries out. Stem spots are dark brown and elliptical. Spots on pods are most common along the pod sutures and are initially dark green and water-soaked, later becoming sunken and dark brown.

Source of infection and spread
Bacterial blight is a seed-transmitted disease. Symptom development in the emerging seedling is favoured if soils and weather conditions are wet. The bacterium can also survive in infected crop debris and spreads in wind-blown rain and on machinery, animals and people moving through a crop when it is wet with dew or rain. The disease can spread rapidly if there is crop injury due to driving rain, hail, wind-blown soil, frosts or insects. The bacterium has only a limited ability to survive in soil in the absence of a susceptible host. Intact crop residues or infected volunteer plants are normally required for the pathogen to survive between crops.

Importance
Bacterial blight of pea has a worldwide distribution but its spread is restricted by quarantine and seed certification. The disease can cause serious losses in southern states of Australia, particularly in field pea.

Management
- Plant disease-free seed.
- Use resistant varieties.

FUNGI

■ BLACK ROOT ROT

Cause
The oomycete *Aphanomyces euteiches*.

Symptoms
Symptoms are not usually visible until a few weeks after planting. Affected plants lack vigour and their root systems and lower stems are discoloured light brown. The colour later changes to a dark brown or black and the outer root tissues become soft. Leaves wilt and shrivel progressively

Fig 18.1 Bacterial blight of peas.

Fig 18.2 Black root rot showing an infected root system (right) and a healthy root system (left).

Fig 18.3 Grey mould on pea pods.

from the base. Provided moisture is adequate, plants do not usually die before setting a reduced crop. The disease is often distributed uniformly through a field and most plants are affected. There is no ready comparison between healthy and diseased plants, so the symptoms are often misdiagnosed as moisture stress.

Source of infection and spread

The pathogen is a soil inhabitant favoured by wet conditions. There are two spore types. A resting spore (oospore) develops in diseased tissue and survives in the soil for several years. If the spores spread with crop residues or soil, the affected area will increase. A second spore type (zoospore) develops during the growing season and travels in water, inecting plants through the roots.

Importance

Black root rot can be widespread and severe in Australian pea crops.

Management

- Avoid poorly drained areas.

- Use long rotations between pea crops.

■ GREY MOULD

Cause

The fungus *Botrytis cinerea*.

Symptoms

Pale, water-soaked spots develop where fallen petals stick to leaves, stems or pods. The spots develop into grey-brown lesions. Late in the season, senescing leaves may become infected during periods of high humidity. Once established

in leaves or stipules, the pathogen can then spread to stems, which wilt and die when lesions girdle the stem. Infected pods develop brown-grey large circular water-soaked lesions. Symptoms may develop on pods post-harvest.

Source of infection and spread

The fungus is a common saprophyte of many host plants that colonies plant debris. Infection occurs after wet weather particularly when temperatures range between 16–21°C. The fungus establishes on the flowers and then spreads to infect the developing pods.

Importance

Grey mould is a minor disease of field, snap and snow peas in Australia.

Management

- Avoid poorly drained areas.

- Use long rotations between pea crops.

- Apply the recommended protectant and eradicant fungicides. Base the spray program on regular protectant fungicide applications, using eradicant products only when environmental conditions are highly favourable for infection, or at the stage of early symptom development.

■ DOWNY MILDEW

Cause

The oomycete *Peronospora viciae*.

Symptoms

Seedlings infected soon after emergence are stunted, and may die. The fungus may invade the whole plant or cause

Fig 18.4 A local infection of downy mildew in pea.

local infection on leaves and pods. Local infection causes indistinct, yellow to brown spots on leaves with a thick grey and brown fungal growth developing on the under surface following moist conditions. Systemic infection leads to stunted plants with severely deformed leaves, which also develop a profuse, grey-brown fungal growth on the under surface.

Fig 18.5 Downy mildew systemic infection in pea.

Source of infection and spread

The fungus produces two spore types. One is a resting spore (oospore) which is embedded in infected plant material and can survive in soil for long periods (up to 10 years), providing an inoculum source for future crops. The other is a wind-dispersed spore (conidium) produced in large numbers during periods of high humidity. This spore forms part of the fungal growth on affected tissues. Severe outbreaks of downy mildew depend on cool (8–20°C), showery weather, whereas dry, warm weather restricts the disease. Leaves become less susceptible to infection as they age.

Importance

Downy mildew occurs widely and commonly affects seedling leaves. The disease is not usually severe on mature plants, except for those with systemic infection.

Management

- Use crop rotation to reduce levels of soil-borne inoculum.
- Spray with the recommended fungicides. Base the spray program on regular applications (seven to 10 day intervals) of protectant products, and apply systemic sprays following periods of cool, wet weather.

■ FUSARIUM WILT

Cause

The fungus *Fusarium oxysporum* f.sp. *pisi*.

Symptoms

Leaflets and stipules turn pale yellow, roll downwards at the margins and wilt. Symptoms usually progress from the

Fig 18.6 Fusarium wilt of snow pea.

Fig 18.7 A pea plant infected with Fusarium wilt alongside a healthy plant.

base of the stem to the apex of the plant. The water-conducting tissues in the stems develop a red-brown discolouration that may extend from the plant crown into the upper stem tissues. Severely-affected plants are stunted, and may die prematurely.

Source of infection and spread
The fungus is a soil inhabitant and spreads with the movement of infested soil and crop residues. The disease develops rapidly when temperatures exceed 20°C. The fungal spores can survive in soils for more than 10 years. The pathogen also spreads with contaminated seed.

Importance
Fusarium wilt may be important in crops grown on light soils during warm weather. It is particularly a problem in snow pea, for which there are currently no resistant varieties available.

Fig 18.8 Discolouration of vascular tissue caused by Fusarium wilt.

Management
• Plant resistant cultivars. There are no resistant varieties of snow pea. Rotate peas with other crops.

• Early planting may enable pods to mature before soil temperatures are favourable for wilt development.

■ LEAF, STEM AND POD BLIGHT

Cause
The fungi *Ascochyta pisi* and *Mycosphaerella pinodes*.

Symptoms
These two related fungi cause spots on leaves, stems and pods. Lesions caused by *Ascochyta pisi* are generally light in colour with a distinct darker margin. Those caused by *Mycosphaerella pinodes* are brown to purple with indefinite margins. The size of lesions varies considerably with weather conditions, ranging from small flecks to stem lesions several centimetres in length. Blight can be very severe in crops with a dense canopy that provides high humidity suitable for disease development.

Fig 18.9 Leaf and stem blight on pea.

Fig 18.10 Leaf spots caused by *Mycosphaerella* on snow pea.

Source of infection and spread

The fungi may be carried on the seed or survive on residues from previous pea crops. Spores may be carried long distances by wind, with severe disease outbreaks occurring following showery weather.

Importance

The diseases occur widely but are not usually severe. Snow pea seems particularly susceptible.

Management

- Plough in crop residues promptly after harvest and rotate crops to avoid disease carryover in the soil.

- Treat seed with a recommended fungicide before planting.

■ POWDERY MILDEW

Cause

The fungus *Erysiphe pisi*.

Symptoms

The disease generally occurs first on the lower leaves. The early stage is difficult to see, since the fungus grows on the surface of the leaves without killing the leaf tissue. Infected portions of the leaf are slightly off-colour, and if the leaf is held to catch the light, the white thread-like growth of the fungus may be seen on the surface. In later stages of development, the disease is easily recognised by the white powdery growth that covers affected areas. Advanced symptoms also include yellow or purple discolouration of the leaf tissues beneath the powdery mildew colonies, early

Fig 18.11 Powdery mildew on pea.

death of severely infected leaves and blight-like symptoms due to foliage desiccation. Pod infection causes seeds to discolour grey-brown.

Source of infection and spread

Powdery mildew survives by growing on a succession of crops or volunteer pea plants and on infected plant debris. There is

Fig 18.12 Leaves affected by powdery mildew.

Fig 18.13 A pea pod infected with powdery mildew compared with a healthy one.

also the possibility of seed-borne transmission. Once established, the fungus spreads rapidly via the large numbers of air-borne spores produced on infected plant parts.

Warm, dry days with cool (optimum 20°C), dewy nights are most favourable for disease development and water-stressed crops seem to succumb to the disease more rapidly. Rainfall is unfavourable for epidemic development.

Importance

Powdery mildew is major problem in areas with long-growing seasons and moist weather in spring. The disease occurs regularly in south-eastern South Australia, Victoria and parts of New South Wales, and it occurs sporadically in drier, shorter season pea growing regions.

Management

- Generally, early season plantings are affected less than later-grown crops or late-maturing cultivars.

- Plant resistant varieties. The high yielding resistant cultivars Mukta and Kiley have superseded the susceptible Glenroy in disease prone areas.

- Spray with the recommended fungicides.

■ FURTHER INFORMATION

Kraft JM and Pfleger FL (Eds) (2001). *Compendium of pea diseases and pests* (2nd edn). APS Press, St Paul, Minnesota.

Potato (*Solanum tuberosum*) is a member of family Solanaceae. Potatoes originated in the Andean Mountain range of South America where they served as a staple in the diet of native people for millennia. Introduced into Europe in the 16th century by the Spanish, it had become an important food crop by the 18th century. The tragic events surrounding the Irish potato famine in the middle of the 19th century means that the disease late blight will always be part of the history of potato culture. The potato is now the third largest food crop in the world, exceeded only by wheat and rice.

Potatoes came to Australia with the early European settlers and soon became an important vegetable crop. Today, potatoes are grown on approximately 35 500 hectares that produce 1.3 million tonnes from about 12 varieties. Potatoes are grown in all States with South Australia, Victoria and Tasmania the largest producers.

Potatoes are propagated by planting tubers or 'seed potatoes,' and the production of healthy, certified seed potatoes is a vital part of a successful industry.

BACTERIA AND PHYTOPLASMAS

BACTERIAL RING ROT – BIOSECURITY THREAT

Cause
The bacterium *Clavibacter michiganensis* subsp. *sepedonicus*.

Symptoms
The pathogen produces both foliar and tuber symptoms. Initially, the lower leaves wilt, particularly during the hottest periods of the day. As the disease progresses the entire plant may wilt, the leaves develop an interveinal chlorosis, and the leaf margins become brittle and necrotic. Entire stems may collapse and die. A milky

Fig 19.1 Bacterial ring rot in a potato tuber.

Fig 19.2 Leaf symptoms of bacterial ring rot.

exudate will ooze from the vascular tissues of the lower stem if a cut is made near the plant crown. In tubers, the vascular ring will develop a yellow to light brown rot, from which a cheesy exudate can be squeezed if tuber symptoms are sufficiently advanced. Eventually, affected tubers will disintegrate completely in the field or post-harvest (if post-harvest storage temperatures exceed 10°C). Ring rot tuber symptoms may resemble those of bacterial wilt and masked by secondary infections of the tubers by opportunistic organisms. In contrast to bacterial wilt, bacterial slime does not ooze spontaneously from the vascular ring without squeezing the tuber, tuber eyes do not die prematurely, bacterial slime does not ooze from the eyes and therefore no 'dirty' eye symptoms (soil adhering to the eyes) which is characteristic of bacterial wilt.

Source of infection and spread

Infected seed potato tubers, which may be symptomless, are the most likely means of spread. Infection occurs readily if contaminated tubers are planted into cool, wet soils. The bacterium transmits readily between infected tubers and healthy seed pieces during seed cutting, and survives for up to five years on the surfaces of crates, bins, sacks and harvesting and grading machinery, even if temperatures drop below freezing. The bacterium also spreads on chewing insects and in irrigation, and it can survive between crops on volunteer potato plants. During the growing season, symptom development is more rapid in dry soils when air temperatures do not exceed 32°C.

Importance

Ring rot is a major disease of potatoes and one of the most important reasons for seedlot rejection in potato certification schemes in North America. The disease is also present in Canada, many countries in Europe and it is widespread in Asia. The disease has never been reported in Australia.

What to do if you suspect bacterial ring rot

This pathogen is a biosecurity risk to Australia. Any suspected affected plants should be reported to the nearest Department of Primary Industries or the Plant Health Australia hotline (1800 084 881).

■ BACTERIAL WILT (BROWN ROT)

Cause

The bacterium *Ralstonia solanacearum*.

Symptoms

Leaves on affected plants wilt suddenly, generally without preliminary yellowing. Cutting the stem at ground level reveals whitish exudate on the cut surface. Leaf wilting and stem collapse may be particularly severe in young succulent plants, especially when the tuber 'seed' is infected. A brown discolouration develops in the vascular tissues of the stem.

On tubers, a wet breakdown occurs at the point of attachment to the stolon and at the eyes. Glistening droplets of dirty white bacterial slime ooze from the vascular ring spontaneously, without sqeezing, when an affected tuber is cut in half. In a more advanced rot, the vascular ring becomes light brown and eventually darkens. The tuber eyes die prematurely and the bacterial slime exudes from the eyes, resulting in a 'milky eye, 'dirty eye' or 'sore eye' symptom. The presence of this ooze distinguishes bacterial wilt from wilts caused by fungi. The rot affecting tubers is often termed brown rot. Infected tubers are prone to invasion by other secondary bacteria, especially *Clostridium* spp., that hasten tuber breakdown.

Fig 19.3 Field symptoms of bacterial wilt.

Fig 19.4 Bacterial wilt ooze symptoms on a potato stem.

Fig 19.6 Internal symptoms of bacterial wilt. Note the beads of bacterial exudate produced when pressure is applied to an infected tuber.

The vascular symptomsin tubers affected with brown rot can be easily confused with those of ring rot.

Source of infection and spread

The bacterium is capable of surviving in soil for many years in the presence of a suitable host. In the absence of a host, the bacterium can survive in soil for up to three years. Commonly introduced in infected 'seed' tubers, bacterial wilt also spreads in infested soil, on machinery and in contaminated water. Tubers washed in contaminated water prior to storage over summer for an autumn planting may become infected with the bacteria. The bacteria infect the root and become systemic in the vascular system of the plant, infecting tubers via the stolons. Plant-to-plant spread occurs through root contact between neighbouring plants.

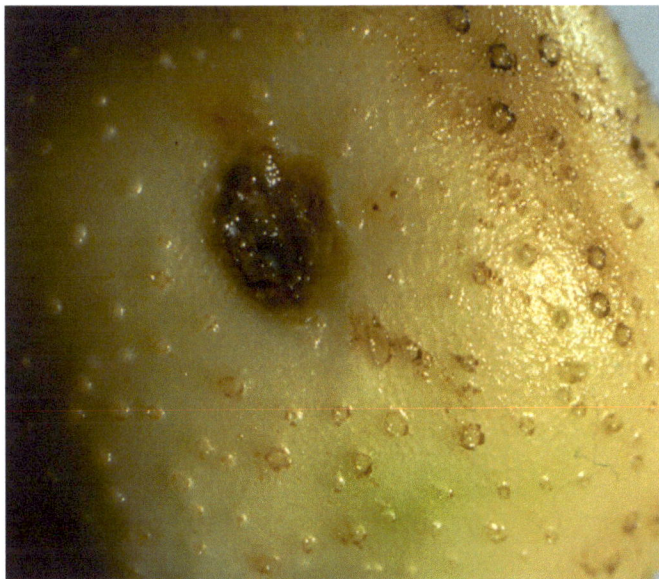

Fig 19.5 Bacterial wilt symptoms on the eye of a tuber.

This disease rarely occurs when the average soil temperature falls below 15°C. When crops mature under cool conditions, latent infection can occur without symptoms developing in the plants or tubers. These tubers are a ready source of infection in the new crop when they are planted as seed potatoes.

Ralstonia solanacearum is a complex species with considerable diversity. There are two forms of bacterial wilt in Australia, caused by two different biovars: biovars 2 and 3. Biovar 2 is specific to potatoes. It is tuber-borne and has been very widely distributed in planting material. Disease caused by this biovar is favoured by cooler temperatures and so this form is most problematic in temperate production areas.

Biovar 3 is common in the high rainfall areas of northern Australia and has a very wide range of hosts among crop and weed species. Many susceptible species are members of the family Solanaceae.

Importance

Bacterial wilt is usually a minor disease on potato, but serious outbreaks occasionally occur, particularly in highly susceptible varieties when favourable soil temperature and moisture conditions prevail.

Management

- Use certified seed.

- Do not source seed potatoes from areas known to have bacterial wilt.

- Grow crops that are not hosts of the pathogen for at least two, and preferably five years before replanting potatoes. Suitable crops include soybean, maize, sorghum and winter cereals.

- Avoid highly susceptible varieties.

- Chlorinate water used in washing plants.

- Grade seed tubers before washing.

- Minimise the movement of farm machinery from affected areas, and clean it before transporting.

■ BACTERIAL SOFT ROT, BLACK LEG AND AERIAL STEM ROT

Cause
The bacteria *Erwinia* spp.

Symptoms
These diseases are very closely related, but black leg and aerial stem rot affect aerial plant organs during the growing season, whereas soft rot is a breakdown of tubers in the field and post-harvest.

Fig 19.7 Black leg symptoms on a potato stem.

Fig 19.8 External symptoms of black leg on tubers.

Black leg describes the characteristic inky black, slimy rot of the lower stems of the potato.

Black leg symptoms may occur in the field at any stage of plant development. Plants appear stunted and often wilt before flowering. The disease initiates from contaminated seed pieces. The seed pieces rot and stems develop a black, wet rot that often extends above ground level. The stem above the rot is greyish-brown to black, and the discolouration not confined to the water conducting tissues. Affected plants die. When pulled by hand, the affected stem easily separates near ground level, leaving 1–2 cm of stronger, stringy plant tissue on the end.

In contrast to black leg, aerial stem rot develops from infection by bacteria external to the seed piece. Symptoms are usually confined to above-ground plant portions. Diseased stems and petioles develop light to medium brown water-soaked lesions that can girdle the stems, causing premature plant death.

Soft rot in tubers develops as a brown, wet, rotted area at the point of attachment, around lenticels or where damage has occurred. Internally, affected tissues are soft and discoloured with a black margin extending inwards from the point of entry.

Tubers can be affected by soft rot while in storage or in the soil before harvest. Soft-rotted tissues are wet, and cream to tan in colour with a soft, granular consistency and do not have an odour. An offensive smell develops as rotting progresses due to the invasion by secondary soft rotting bacteria.

Source of infection and spread
The bacteria are common soil inhabitants and are often dispersed in infected seed tubers and in irrigation water. In black leg, infection of the plant can occur through the base after the seed piece breaks down, or through stem and leaf wounds. Seed piece breakdown is most likely to develop in wet soils at temperatures above 20°C. When a film of water covers a seed piece, oxygen depletes rapidly and the seed piece decays quickly. Bacteria in the decaying seed pieces

may then move through the water-conducting tissues into the stems, where stem lesions may develop. The bacteria also move through the roots to infect the young tubers. This leads to a soft rotting of tubers that is commonly visible as a post-harvest problem. In addition, *Erwinia* enters tubers through lenticels, cuts and others wounds. Soft rot is often serious where tubers have been washed in contaminated washing plants.

In aerial stem rot, above-ground stems and petioles become infected when the pathogen enters wounds caused by hail, wind-blown sand, insect feeding or cultivation. Natural openings such as leaf scars or stem cracks provide alternative sites for the development of aerial stem rot.

Importance
Black leg and soft rot are important diseases of potatoes that can result in a significant reduction in quality. Some consignments of seed tubers can have high levels of infection. Aerial stem rot is a less common symptom, although it may develop in cool, wet conditions.

Three species of *Erwinia* can be involved: *Erwinia carotovora* subsp. *carotovora*, *E. carotovora* subsp. *atroseptica*, and *E. chrysanthemi*. *E. chrysanthemi* is the most common in warm climates.

Management
- Use certified seed.

- When using cut seed, thoroughly clean all seed cutting and handling equipment and treat with a sanitiser between seed lots.

- Minimise bruising of seed tubers by lowering conveyors that move seed tubers onto trucks and planters, to reduce the distance the seed pieces drop.

- Plant well-suberised seed pieces into well-drained cool soils (10–13°C at the planting depth).

- Calibrate harvesting and tuber-handling equipment so that mechanical damage to tubers and bruising is minimised.

- Use only clean water to wash potatoes.

- Harvest tubers when soil temperatures are less than 20–25°C and when tubers remain in the ground for 7–10 days after plant death. Tubers harvested from green plants are more susceptible to post-harvest soft rot.

- Do not wash tubers before storage.

- Maintain stored tubers at 10–13°C and 95% relative humidity for 10–14 days to promote wound healing. After this period, tubers should be stored below 10°C in well-ventilated conditions.

■ COMMON SCAB

Cause
The bacterium *Streptomyces scabiei* (syn. *S. scabies*).

Symptoms
Affected tubers have circular, corky, deeply pitted areas usually 5–8 mm wide but often coalescing. Less commonly, small russet areas or slightly raised lesions develop on tubers.

Source of infection and spread
The bacterium is common in potato-growing soils where it may survive for many years on decaying organic matter. Introduced into clean areas on infected seed tubers, it spreads when infested soil is moved. Infection occurs through immature lenticels and tubers are most susceptible for six to eight weeks following tuber initiation. The optimum temperature for growth of *S. scabiei* is 30°C and infection is usually more severe in dry soils with a pH of 5.2 to 7.0.

Importance
The disease is common in Australia. It generally does not affect yield but it can reduce tuber quality sufficiently to reduce tuber grade. The disease is generally not serious where water is adequate after tuber set and during the early enlargement phase.

Management
- Do not plant diseased seed.

- Apply adequate irrigation to maintain soils close to field capacity for four to six weeks, beginning at tuber initiation.

- Monitor soil moisture levels with tensiometers.

Fig 19.9 Severe common scab on tubers.

Fig 19.10 A range of common scab symptoms on tubers.

- Apply the registered fungicide seed treatments to reduce the risk of seed-borne scab.

- In soils with a high incidence of disease, a three to four year rotation with lucerne or soybeans will help to reduce the soil population.

- Apply acid fertilisers such as ammonium sulphate to help reduce soil pH to less than pH 5.5.

■ PURPLE TOP WILT

Cause
A phytoplasma.

Symptoms
Initially, young leaves roll upwards, generally at flowering. Affected plants become stiff and erect and develop yellow or purple discolouration at the base of young leaflets. Internodes are shortened, giving a rosette effect. Aerial tubers may form in the leaf axils. Seed pieces of affected plants generally rot and water-conducting tissues are discoloured. Tubers may be flabby and fail to sprout, or may produce spindle sprouts that do not develop into plants.

Source of infection and spread
Purple top wilt (sometimes called autumn wilt) is most serious in autumn plantings. Phytoplasma spread by leafhoppers from some weeds, are a likely cause of purple top wilt disease.

Importance
Purple top wilt can cause moderate losses in the autumn-planted crop.

Management
There is no control available for this disease.

FUNGI

■ BLACK DOT (ANTHRACNOSE)

Cause
The fungus *Colletotrichum coccodes*.

Symptoms
The name black dot describes the abundant, dot-like black sclerotia on affected tubers, stolons and stems. Above ground, plants slowly wilt and older leaves yellow. Young leaves may show vein death, interveinal scorching and rolling of the leaf margin. Below ground, the lower stem may be dry and shredded with numerous black sclerotia embedded in the tissues. Sclerotia can develop on the upper surface of tubers. The infected skin on a tuber turns grey during storage resulting in symptoms that somewhat resemble silver scurf.

Fig 19.11 Purple top wilt.

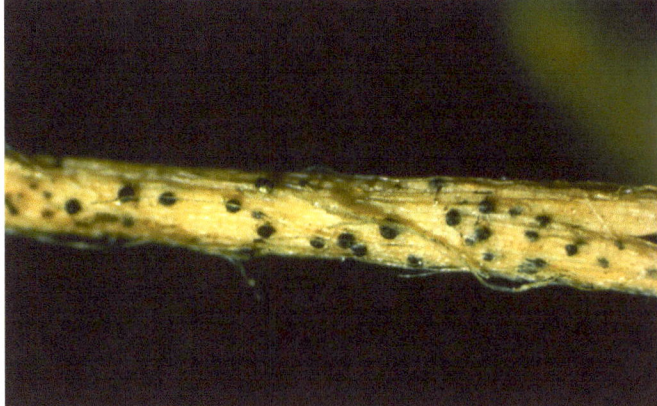

Fig 19.12 Black dot showing black, 'seed-like' fruiting bodies (sclerotia) in infected potato stems.

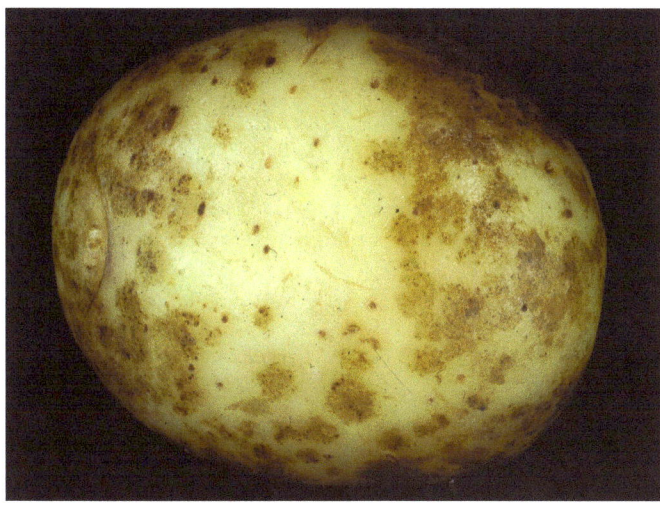

Fig 19.13 Black dot symptoms on a tuber.

Source of infection and spread

The fungus survives as sclerotia in the soil, in crop debris, on volunteer potato plants and on the surface of tubers. It can also infect other members of the *Solanaceae* plant family. The fungus gradually colonises the underground parts of potato plants as the crop develops. Infected seed tubers can introduce the fungus into new areas. Black dot is usually more prevalent under warm, dry conditions.

Importance

Black dot is commonly found in most potato growing regions. The disease causes surface blemishing of tuber skins resulting in downgrading in potatoes. It has become more noticeable since it has become more common to wash fresh market potatoes. It often occurs in combination with other soil-borne pathogens including *Fusarium, Rhizoctonia* and *Verticillium* species.

Management

- Plant certified seed potatoes.

- Avoid growing potatoes in the same ground more than once in every three years.

- Ensure good nutrition and irrigation of the crop, particularly in sandy soils.

■ DRY ROT

Cause

The fungi *Fusarium* spp. and *Phoma* sp.

Symptoms

Fusarium rot: This is a dry, sunken rot, often giving tubers a wrinkled appearance. Pockets within the rot generally fill

Fig 19.14 Dry rot symptoms caused by *Fusarium* on tubers.

with white or pink wefts of the fungus that may sometimes be present on the outer surface. If the rot starts around the point of attachment, it is known as 'stem end rot' and may indicate the presence of Fusarium wilt in the parent crop. Other fungi can also cause stem end rot.

Phoma rot: Shallow, well-defined, thumb-mark depressions with small, pinpoint, black dots (fruiting bodies) occur on the tuber surface.

Source of infection and spread

The fungi are common soil inhabitants that enter the tuber through wounds. Because the rot develops slowly, it is generally visible only after potatoes have been stored for some time. Relatively high storage temperatures favour the development of Fusarium rots but low temperatures favour the development of Phoma rots. Planting affected seed usually results in a poor stand.

Importance

Fusarium dry rots of tubers can occur in all cropping areas and can affect tubers in storage as well as cut surfaces of seed setts before and after planting. Injury to tubers during

Fig 19.15 Dry rot symptoms caused by *Phoma* on a tuber.

harvesting and handling is a pre-requisite for storage rots followed by poor storage conditions.

Management

- Inspect seed consignments on arrival and reject those affected.

- Store the consignments in cool, well-ventilated conditions.

- Reject affected tubers before cutting.

- Apply the recommended fungicides to tubers before storing.

- Avoid injury to tuber skins by not harvesting under very cold or very dry conditions.

- Optimise grading equipment and grading speed to minimise damage to tuber skins.

- After harvest or grading, allow a curing period (above 12°C with good ventilation for at least one week) before placing tubers in a cool store.

■ FUSARIUM WILT AND STEM END ROT

Cause
The fungi *Fusarium oxysporum* and *F. solani*.

Symptoms
Affected plants turn yellow, rapidly wilt and die. Initially, symptoms may be confined to one stem of the plant. The seed piece decays and the lower stems develop a dry rot.

The water-conducting tissues of the stem may be discoloured brown.

The fungi can progress along the stolons to infect young, developing tubers and cause stem end rot. Water-conducting tissues in these tubers usually discolour brown. The symptoms resemble those of potato early dying (caused by *Verticillium*). Stem end rots are sometimes associated with a stressful finish to the crop, for example hot, dry conditions.

Source of infection and spread
The fungi are common soil inhabitants and introduced also on infected seed. Symptoms are most apparent at high soil temperatures and low soil moisture.

Importance
Fusarium wilt and stem end rot can cause important crop losses, particularly in warmer production areas.

Management
- Plant certified seed.

Fig 19.16 Fusarium wilt.

- Before planting, remove tubers showing symptoms of dry rot.

- Long rotations out of potatoes help to reduce soil-borne inoculum.

■ LATE BLIGHT (IRISH BLIGHT)

Cause
The oomycete *Phytophthora infestans*.

Symptoms
Leaf, stems and tubers of the potato plant are susceptible to infection by *Phytophthora infestans*. The appearance of leaf lesions varies depending on their age and the ambient environmental conditions. Initially, leaf lesions are small, dark and irregular in shape. When conditions are favourable, lesions become large, pale green areas with indefinite margins, becoming water-soaked and dark in

Fig 19.17 Leaf symptoms of late blight on potato.

colour. Whitish, downy 'fuzz' may be visible on lesion margins on the underside of the leaf and surrounding tissues in moist weather.

Stem lesions commonly occur at the apex of shoots or at stem-petiole junctions. Stem lesions are initially water-soaked, but may dry out with time in low humidity.

Fig 19.18 Late blight symptoms on the underside of an affected potato leaf.

Fig 19.19 Late blight lesions occurring at stem-petiole junctions in potato.

Slightly sunken areas of dark, rotted tissue may occur on tubers. Tuber lesions are generally firm, although secondary infection by other organisms may cause tubers to break down rapidly.

Source of infection and spread

The pathogen survives on potato residues and volunteer potatoes. Large numbers of spores develop in the downy 'fuzz' on affected areas and these spread in wind, rain or irrigation water. Cool (18–22°C), wet weather favours infection and disease development. Suitable weather conditions are those that allow moisture (dew, heavy mist, irigation or rain) to remain on the leaf and stem for prolonged periods. Under optimal conditions, leaf lesions can develop within six days of infection.

Importance

Although an important disease of potatoes worldwide, late blight is of minor importance in most years in Australia. Weather conditions are seldom favourable for the disease. Australia still only has the archetypal A1 mating type population of *P. infestans*. In most of the rest of the world, this A1 mating type population was replaced in the last two decades by more aggressive strains of a *P. infestans* population consisting of both the A1 and A2 mating type. These are more virulent, causing a more severe disease. They have much shorter life cycles, cause serious stem lesions, are resistant to phenylamide fungicides (for example, metalaxyl) and are active under broader environmental conditions than the old strains. It is important to safeguard Australia from the introduction of this more aggressive *P. infestans* population. Most potato varieties are very susceptible to late blight.

Management

- Maintain a regular protectant fungicide spray program and include registered systemic fungicides when conditions favour infection, i.e. after cool, wet weather.

- Avoid highly susceptible varieties.

■ PINK ROT

Cause

The oomycete *Phytophthora erythroseptica* is the main cause of the disease.

Symptoms

Tuber infection often begins at the stem end, and the affected area is frequently bordered by a dark line visible through the skin. These borders are easily seen once soil is washed from the tubers. Buds and lenticels are black and often produce a clear liquid. Infected tubers are rubbery but not soft and infected skin is easily rubbed from the tuber. When cut and exposed to air, recently infected surfaces change from cream coloured to salmon-pink in about 30 minutes, and then to black after about one hour. This colour change is often associated with a characteristic pungent smell of ammonia, which may be evident in storage areas before symptoms are apparent in tubers.

Tubers affected by pink rot are often invaded by soft rot bacteria causing slimy, tuber decay.

The only above-ground symptom of pink rot is a late-season wilt when plants appear limp and flat compared with healthy plants.

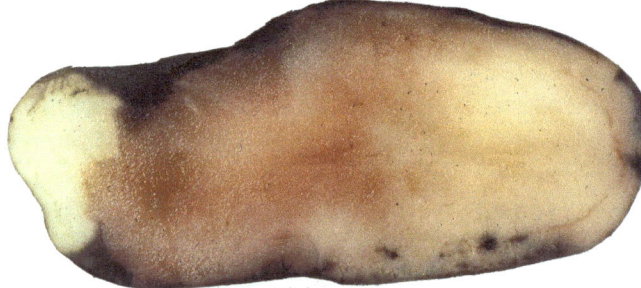

Fig 19.20 Pink rot symptoms on a tuber.

Source of infection and spread

The pathogen has a wide host range and can persist in the soil for several years, usually as thick-walled oospores.

The pathogen can be introduced in infested soil on bulk bins and equipment, by drainage water from infested areas and as a contaminant on the surface of sound tubers used for planting.

All underground parts of the plant are susceptible and infection occurs from sporangia produced from oospores or zoospores. Tubers are generally infected via infected stolons but infection can also occur through eyes and wounds.

Infection and disease development is favoured by warm temperatures (15–25°C) and wet soils. High levels of pink rot are most likely to occur in low lying waterlogged areas or following periods of very wet weather.

The spread of pink rot in storage is generally due to the development of field infections rather than spread of the pathogen in storage areas.

Importance

Pink rot can cause serious losses in stored tubers grown in very wet soil.

Management

- Apply registered fungicides as per label recommendation.

- Plant in well-drained soils that do not have a previous history of the disease.

- Avoid excessive irrigation, particularly late in the growing season.

- Avoid wounding tubers during harvest.

- If pink rot is found in a field, harvest late to allow time for affected tubers to rot in the field.

- Maintain good airflow and low temperature in storage, and avoid the accumulation of moisture on tubers.

- Rotate potatoes with other crops to reduce soil inoculum levels.

■ POTATO EARLY DYING

Cause

In Australia, the fungus *Verticillium dahliae* is the primary cause of the disease.

Symptoms

Affected plants wilt slowly and die prematurely. A close examination of plants shows death of the older leaves and

Fig 19.21 Potato early dying symptoms on a whole plant and in a stem section.

yellowing of the middle leaves. One side of the plant may be affected more severely than the other. A distinguishing feature of the disease is a golden to brownish yellow discolouration of the water-conducting tissues of the stem. This is best seen when the stem is cut across in a long slant at about ground level. Affected tubers show a dark discolouration of the water-conducting tissues, particularly near the point of stolon attachment.

Source of infection and spread

The fungus is a soil inhabitant, occurring naturally in many soils and carried in or on seed potatoes. Once established in a field, the fungus can persist for many years as microsclerotia. It can persist in the roots of many symptomless crops and weed species, as well as other susceptible hosts including tomato, cotton and Noogoora burr (*Xanthium pungens*).

Importance

Verticillium wilt is more serious in autumn crops and often limits crop production where rotations are not possible.

Management

- Do not plant affected seed.

- In soils with high levels of the fungus, rotate to cereals or green manure crops to reduce levels of soil inoculum. Do not grow potatoes more than once in three years and eliminate volunteer potato plants.

- Apply a registered soil fumigant to fields with high levels of inoculum.

POTATO SMUT – BIOSECURITY THREAT

Cause

The fungus *Thecaphora solani* (syn. *Angiosurus solani*).

Symptoms

No symptoms develop on above-ground plant parts. Spherical to irregular galls form on subterranean stems, stolons and tubers. The galls vary in size and frequently have a scaly or cracked outer layer. On stems and stolons, galls appear initially as small lateral outgrowths that increase in size and surround the organs to which they are attached. The biggest galls form on the underground stems, and those on the stolons tend to be smaller. On tubers, two types of symptoms develop: small galls (most commonly at the apical ends of the tubers); and slight protuberances that become sunken and suberised post-harvest. Although no galls form on the plant roots, roots may often arise from

Fig 19.22 Potato smut on tubers and a stem.

the galls. Dark brown spore masses arranged in a roughly radial pattern within the galls may be visible if a transverse cut is made through the gall tissues.

Source of infection and spread
Infected seed potato tubers are the most likely means of spread. The fungus may persist in soils for many years in gall fragments and tuber debris. Galls are frequently observed 45–60 days after planting.

The pathogen can develop in different soils and climates, including cool, mountaineous regions and warmer coastal climates in subtropical areas.

The main hosts are potato and other species of *Solanum*.

Importance
Potato smut is indigenous to the Andean region of South America but also found in Panama and Mexico. The disease is serious, with production losses of up to 80% reported in susceptible cultivars. The pathogen has not been recorded in Australia.

What to do if you suspect potato smut
This pathogen is a biosecurity risk to Australia. Any suspected affected plants should be reported to the nearest Department of Primary Industries or the Plant Health Australia hotline (1800 084 881).

POTATO WART – BIOSECURITY THREAT

Cause
The fungus *Synchytium endobioticum*.

Symptoms
The characteristic symptom is the production of galls on several plant parts, particularly the tubers and stolons (underground stems). Galls occasionally develop on the upper stem, leaf or flower. On infected tubers, the eyes develop into characteristic, warty, cauliflower-like swellings. Galls formed below ground are white to brown and turn black as they decay. Above-ground galls are green to brown, turning black at maturity and later decaying. Tubers may be disfigured or become completely replaced by galls.

Symptoms of wart disease can be confused with those of powdery scab and laboratory tests are necessary to identify the pathogens accurately.

Source of infection and spread
The fungus is an obligate soil-borne pathogen naturally infecting only potato. It survives in the soil as resting sporangia for at least 40 years in the absence of the host. Infection occurs as motile zoospores, released by the sporangia and which encyst, causing galls to develop on susceptible host tissue. Water is required for sporangia germination and zoospore movement. Prolonged cool, wet weather favours the disease.

The fungus disperses mainly on infected seed potatoes and contaminated soil on tubers, farm machinery and packing crates.

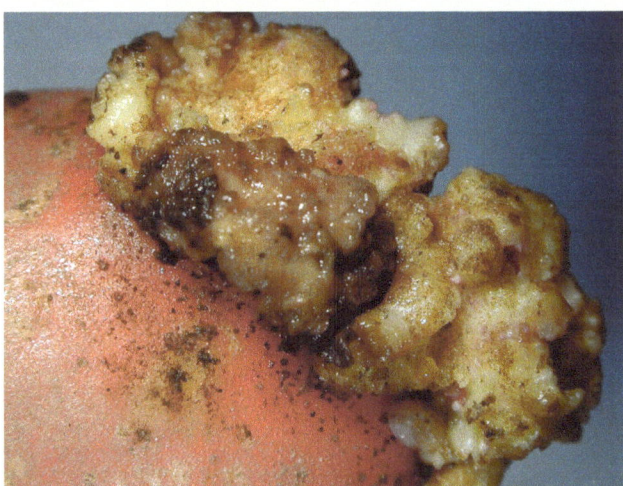
Fig 19.23 Potato wart disease.

Importance
Potato wart was once the most serious disease in potatoes in many countries. Although now generally well managed, it is a biosecurity risk for the Australian potato industry because of the long survival time of spores in contaminated soil. Strains of the fungus able to infect wart-resistant potato varieties have developed in Europe.

The pathogen occurs in limited areas on all continents except Australia.

What to do if you suspect potato wart disease
This pathogen is a biosecurity risk to Australia. Any suspected affected plants should be reported to the nearest Department of Primary Industries or the Plant Health Australia hotline (1800 084 881).

Fig 19.24 Powdery mildew in potato.

POWDERY MILDEW

Cause
The fungus *Golovinomyces cichoracearumi* (syn. *Erysiphe cichoracearum*).

Symptoms
Grey-white powdery patches develop on the leaves, particularly on the undersides. Leaves become chlorotic and then develop dark brown necrotic spots. Elongated, light brown stippled flecks, 0.5–2 mm long, develop on the stems and petioles. Grey-white powdery growth develops on the flecks. With time, the flecks may form larger, water-soaked, blackened spots.

Source of infection and spread
Spores of the fungus spread in wind and may travel at high altitude over long distances. Germination of the spores occurs over a wide range of environmental conditions. Damage is most severe on drought or nutritionally stressed plants. Dry, warm weather favours the development of the disease. Powdery mildew carries over between crops on volunteer potatoes and weed hosts.

Importance
Powdery mildew is a very minor disease of potatoes in Australia.

POWDERY SCAB

Cause
Spongospora subterranea.

Symptoms
Raised, wart-like growths on tubers eventually dry out and develop into powdery 'scabs' or slightly sunken scabs

Fig 19.25 Powdery scab on tuber.

fringed with torn tuber skin. These symptoms can be confused with common scab. Powdery scab can be confirmed when characteristic sporeballs or cystosori are found in the powdery material by microscopic examination.

Powdery scab also occurs on potato roots as small cauliflower-like growths or 'galls' that also contain cystosori.

Source of infection and spread
The pathogen is generally introduced with seed but, once established in the soil, it can survive for at least six years in the absence of potatoes as a mass of thick-walled resting spores (cystosori). In response to the presence of host plants, the cysts germinate, producing zoospores that move in free water and infect the potato roots. The spores encyst and undergo further reproduction. The disease is favoured by cool (11–18°C), moist (periods of saturation) soil conditions. It is often most severe in wet areas of the field, or in crops which have been heavily watered to reduce frost injury. The pathogen generally infects tubers through the lenticels. A critical period is early tuber set-a period of about four weeks when tubers are susceptible to infection. The scabs, however, do not appear on the tuber until the crop begins to mature. The fungus is not active in acid soils with a pH below 4.5.

Importance
Powdery scab is a widespread and serious disease of potatoes in Australia. The disease affects tuber quality and scabby tubers are prone to secondary rots in storage. The organism is also the vector of *Potato mop-top virus* which has not been found in Australia.

Management
- Plant disease-free seed.

- Plant resistant varieties.

- Do not apply lime before growing a winter crop of potatoes in light, acid soils.

- Reduce the frequency and quanity of irrigation during early tuber set on very susceptible varieties.

- Avoid cropping tomato in rotation with potato and control solanaceous weeds such as nightshade.

■ RHIZOCTONIA CANKER (BLACK SCURF)

Cause
The fungus *Rhizoctonia solani*.

Symptoms
Dry, brown, sunken elongated lesions or cankers develop on young potato sprouts below ground level. Lesions may girdle the sprouts completely causing the part above the lesion to die. Stem bases and stolons on the growing plant can also be affected. Damage to the stem is often evident as slightly yellowed, stunted, pinched and rosetted top leaves. Another common symptom is the formation of aerial tubers in the leaf axils. Infection of the stolons causes their tips to die, resulting in fewer tubers or the branching of stolons and a characteristic 'nest' of small tubers clustered near the stem at the soil line.

Small, raised, black fungal bodies (sclerotia) are firmly attached to the surface of the tubers. These are sometimes mistaken for soil particles and are most obvious when the tuber is wet. These develop as tubers mature. Blind eyes, russeting of the skin and surface cracking of the tubers may also be associated with *Rhizoctonia* damage.

Source of infection and spread
The fungus is a common soil inhabitant and is very active in decaying plant material. It may also be introduced with

Fig 19.27 Rhizoctonia rot (black scurf) on a potato tuber.

seed. Persistent, cool, wet weather (soil temperatures around 18°C) is favourable for infection. Young sprouts are particularly susceptible to infection. Delayed harvest of tubers growing in cool, wet soils favours black scurf. There are several strains of the fungus and the potato strain (AG3) is usually common in areas used regularly for potato production.

Importance
Black scurf has become a major problem for washed, fresh market and seed potato growers in Australia. It is particular problem in southern parts of Australia, when weeks of cold weather occur following planting. A high incidence of black scurf can sometimes occur in consignments of seed potatoes.

Management
- Plant certified seed.

- If there is a risk that the seed is contaminated, even with only low levels of *Rhizoctonia*, apply a registered seed treatment.

- Ensure that sprouts emerge quickly after planting so that damage is minimised.

  - Do not plant into cold, wet soils.

  - Warm the seed (to more than 15°C), before planting.

  - Plant no more than 5 cm deep.

  - Ensure seed has sprouted before planting (do not plant dormant seed).

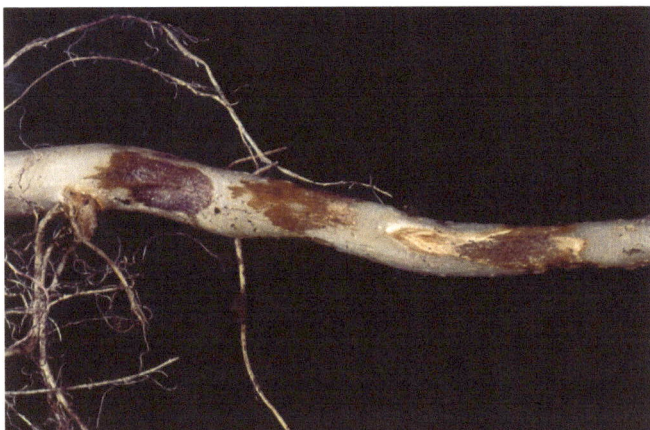

Fig 19.26 Rhizoctonia rot causing dry, brown sunken spots on a young potato stem.

- Prepare land early so that crop residues have enough time to decompose.

- Harvest tubers as soon as possible after vine death.

  - Mechanical vine-kill delays scurf development more than chemical vine-kill.

- Where possible, plant in fields without a history of *Rhizoctonia* and follow-up with a two to three year rotation out of potatoes, or a three to five year rotation if *Rhizoctonia* has proven troublesome in the preceding potato crop.

- Prior cropping with onions, corn, grasses or cereals reduces carry-over of sclerotia.

- Destroy potato volunteers and Solanaceae weeds (e.g. nightshade).

■ SCLEROTINIA ROT

Cause

The fungi *Sclerotinia sclerotiorum* and *S. minor*.

Refer to the chapter on Common diseases of vegetable crops.

Fig 19.28 Sclerotinia rot on potato stems.

■ SEED PIECE BREAKDOWN

Cause

Seed piece breakdown is a physiological disorder.

Symptoms

Seed piece breakdown is characterised by the seed piece rotting, then brown discolouration, followed by the lower stem and roots rotting. Plants may fail to emerge or produce weak shoots, but often there is no evidence of the disease visible above ground.

Source of infection and spread

Seed consignments differ in their susceptibility to the disorder. Susceptibility is affected by the growing conditions of the seed crop, storage conditions, length of storage and handling. Susceptible consignments do not withstand cutting injury and are then easily rotted by soft-rot bacteria. Although affected plants may emerge normally, the rotted seed piece provides easy access to the stem for wilting organisms, particularly *Fusarium* spp. and *Rhizoctonia solani*, which cause wilting and death.

Importance

Seed piece breakdown is an important cause of problems in establishing potato crops.

Management

- Avoid storing seed for long periods.

- Recondition cold-stored seed at normal temperatures for two to three weeks.

- Cut in the shade under cool conditions.

- Treat cut seed with a registered fungicide.

Fig 19.29 Seed piece breakdown in potato.

- Hold treated seed in well-ventilated conditions.

- Plant into moist soil soon after cutting and do not irrigate before emergence.

■ SILVER SCURF

Cause
The fungus *Helminthosporium solani*.

Symptoms
Silver scurf is first seen as small, localised, circular, light brown spots on the tuber skin. As the disease progresses, grey to silvery blotches, roughly circular in shape develop. The entire tuber may eventually be affected. The silvery blotches can look sooty black when the fungus produces an abundance of black spores in warm, humid conditions. The fungus only enters the outermost skin layer (periderm). Diseased areas have a distinct silvery sheen, particularly if the surface is wet. Affected areas may dry and slough off, allowing water loss during storage which results in tubers becoming shrivelled and wrinkled.

Source of infection and spread
Potato is the only known host of *Helminthosporium solani*. The fungus can survive in the soil between seasons or be introduced on seed potatoes. High humidity (more than 90%) is necessary for disease development on the tuber skin, which becomes more severe the longer the tubers remain in the ground. On recently harvested tubers, the disease is more common on the stem end. Silver scurf is often more severe on tubers in storage, particularly towards the end of the storage period. Spores are present in dust and soil in storage facilities and are easily spread by air currents, infecting tubers during periods of condensation on the skin.

Fig 19.31 Silver scurf on tubers.

Importance
Silver scurf is a very common skin blemish disease of potatoes. The disease does not generally affect crop growth or yield, but it reduces the cosmetic quality of washed potatoes. Severely affected tubers can also affect the processing quality of crisps, which may have blackened edges, making the product unmarketable.

Management
- Plant certified seed.

- Apply a registered pre-plant fungicide

- Harvest tubers as soon as possible after vine death.

- Allow tubers to dry as soon as possible after harvest.

- Treat seed potatoes with a registered fungicide seed treatment prior to storage to prevent the fungus from producing spores.

- Remove any soil or debris that could harbour the disease from the tubers before storage.

- Clean and disinfect harvesting equipment frequently.

- Separate the storage of early generation seed from ware potatoes wherever possible.

■ TARGET SPOT (EARLY BLIGHT)

Cause
The fungus *Alternaria solani*.

Symptoms
Early symptoms are small, brown to black spots on lower leaves. These enlarge to 20 mm wide, often with a marked zonate pattern. The lesions are initially circular, but as they enlarge, are frequently delimited by the veins, giving them an angular shape. The disease generally affects crops after flowering towards early senescence. At the early stages of

Fig 19.30 Silver scurf (left) and black dot (right).

Fig 19.32 Target spot.

Fig 19.33 Target spot on a potato leaf.

infection of the crop, whole plants may be 'peppered' with spots. At the later stages of development whole leaves become brown, dry and crispy. The entire plant may be defoliated in a severe infection. Yield loss can occur if the disease becomes serious during the stage when the tuber is bulking up. Plants under stress are more prone to target spot than vigorous, healthy plants, although most crops will eventually develop target spot towards the end of their life.

Dark, circular to irregular, sunken lesions may develop on the tubers. Infected tubers shrivel after extended storage, especially at high temperatures.

Source of infection and spread

The common sources of infection are diseased plants and residues of tomatoes and potatoes. Spores produced on the spots spread in wind, rain or irrigation water. Warm weather (20°C optimum) and high humidity favours the disease. Spores require moisture on the leaf surface for germination, although alternating periods of wetting and drying favour rapid development of target spot. Tuber infection only occurs if contaminated soil contacts wounds made during harvest.

Importance

Target spot is the major leaf disease of potato throughout Australia.

Management

- Use a preventative spray program with recommended fungicides, beginning the applications after row closure.

- Maintain a healthy crop with appropriate nutrition and irrigation to avoid plant stress.

- Mechanical vine removal or chemical vine desiccation and delayed harvest will promote skin set, reduce skinning injury and reduce tuber infection.

NEMATODES

POTATO CYST NEMATODE – BIOSECURITY THREAT

Cause
The nematodes *Globodera rostochiensis* and *Globodera pallida*.

Symptoms
Affected plants are stunted and chlorotic and fail to close the canopy across rows. Plants may wilt during the day. Tubers are smaller but quality is not affected.

The nematode is usually present in a field for about seven years before nematode numbers reach a detectable level.

The distribution of affected plants is usually patchy due to the uneven distribution of the nematodes. The above-ground symptoms of infestation are not distinct and are similar to those associated with many soil-borne pathogens, moisture stress and some nutrient deficiencies.

Fig 19.34 Potato crop affected by potato cyst nematode.

Fig 19.35 Potato cyst nematode on roots. Note the golden, spherical appearance of the encysted female nematodes.

The presence of the nematode is most easily confirmed in the field by the presence of round, pin-head sized white to golden yellow cysts on the roots of affected plants at flowering. The cysts are visible with the naked eye, but are best found with the aid of a hand lens. Cysts dislodge easily from roots, so the roots need to be lifted and handled with care.

Laboratory tests are essential to confirm the presence of the nematode.

Life cycle

In the absence of a host, eggs of both species survive within cysts in soil for periods of up to 30 years. A cyst is a dead body of a female and may contain up to 500 eggs. Eggs hatch when stimulated by exudates from roots of a growing host plant. Juveniles penetrate the roots, locate suitable feeding sites and then undergo three moults during their development to the adult stage. The adult male leaves the root and the female remains in the root feeding and growing. After mating, females fill with eggs. The adult female (the cyst) is initially a creamy-white colour, but becomes tan as roots die. *Globodera rostochiensis* turns a golden colour just before it tans, hence this nematode is also known as the 'golden potato cyst nematode'. The cyst of *Globodera pallida* remains white until it tans (white potato cyst nematode).

The hosts of the potato cyst nematodes are restricted to the Solanaceae or potato family, and include eggplant, tomato and some weed species.

The cysts spread in infested soil on tubers, on farm implements and machinery, boots, transport crates and plant parts. The cysts can also be carried by drainage and flood water, and can be transported by wind.

Importance

The potato cyst nematode originated in the Andes of South America and spread with the transport of potatoes and contaminated soil to many potato-growing regions of the world. Found mainly in temperate regions, it also occurs in coastal areas of the tropics.

The golden potato cyst nematode has a very limited distribution in Australia and is under active control. *Globodera pallida* has not been found in Australia.

The eradication of existing infestations of the nematode and quarantine measures to prevent further spread or introductions are a major biosecurity issue for the Australian potato industry, because of the potential effects on production, and access to export and domestic markets.

What to do if you suspect potato cyst nematode

This pathogen is a biosecurity risk to Australia. Any suspect affected plants should be reported to the nearest Department of Primary Industries or the Plant Health Australia hotline (1800 084 881).

Fig 19.36 Root-knot nematode in potato.

■ ROOT-KNOT NEMATODE

Refer to the chapter on Common diseases of vegetable crops.

VIRUSES

■ CALICO

Cause
Alfalfa mosaic virus.

Refer to the chapter on Common diseases of vegetable crops.

Fig 19.37 Calico disease.

Importance
Calico disease is seldom a problem in potato crops. Infection is most likely to occur when potatoes are grown near legumes such as clovers and lucerne.

■ LEAFROLL

Cause
Potato leafroll virus (Polerovirus).

Symptoms
The development of potato leafroll is often termed primary and secondary infection. Symptoms of primary leafroll develop in plants infected by aphids during the growing season. Younger leaves show rolling first, particularly at the base of the leaflets, and they turn slightly pale in colour. These symptoms may progress to the lower leaves. Secondary leafroll symptoms occur in plants grown from diseased tubers. Infected plants show upward rolling of the leaf margin, starting with the lower leaves and progressing upward. Leaves are harsh in texture. Younger leaves become erect and pale in colour and cut tubers may show an internal net browning or reduce in size.

Source of infection and spread
The virus is carried in tubers and is efficiently transmitted by aphids that breed on potatoes. Important aphid carriers include the green peach aphid (*Myzus persicae*) and the potato aphid (*Macrosiphum euphorbiae*). The virus can survive in aphids for many weeks and spreads over long distances by winged aphids and over short distances within

Fig 19.38 Field symptoms of potato leafroll disease.

Fig 19.39 A potato leafroll affected plant.

a crop by aphids in their wingless stage. Older, infected crops provide a source of infection for young crops. Other hosts of *Potato leafroll virus* are tomato and the weed species thornapple (*Datura stramonium*), apple of Peru (*Nicandra physalodes*), gooseberry (*Physalis* spp.), shepherd's purse (*Capsella bursa-pastoris*) and deadnettle (*Lamium amplexicaule*).

Importance
Leafroll is one of the most serious and common virus diseases in potato in Australia, and is a major disease in both autumn and spring crops in Queensland, with losses exceeding 50% at times.

Management
- Use certified seed potatoes.
- Avoid using seed tubers saved from crops affected by leafroll disease.
- Control aphids with recommended insecticides.
- Avoid growing young crops near older infected potato crops and other virus hosts.

■ POTATO VIRUS X

Cause
Potato virus X (Potexvirus, PVX).

Symptoms
The virus usually produces mild symptoms, though some strains of the virus can cause severe mosaic symptoms, and

tuber necrosis may occur in some varieties. If diseased seed is planted, yields may be reduced, and the combination of this virus with *Potato virus Y* produces rugose mosaic, which may cause serious losses.

Source of infection and spread
The virus is transmitted on tubers of infected plants. It is also very infectious and spreads by contact between diseased and healthy plants, on hands and equipment and during cutting of seed pieces.

Importance
The virus is rare in certified seed potatoes and uncommon in commercial production in Australia.

Management
- Use certified seed potatoes.
- Avoid mechanical damage to plants during crop maintenance.
- Sanitise equipment, especially during seed piece preparation.

■ POTATO VIRUS Y

Cause
Potato virus Y (Potyvirus, PVY).

Symptoms
Symptoms are extremely variable, ranging from mild mosaic and vein clearing symptoms, crinkling and wavy

Fig 19.40 *Potato virus Y* on a potato tuber.

Fig 19.41 *Potato virus Y* on a potato leaf.

leaf margins, through to severe mosaic, necrosis and leaf drop. Some strains in certain cultivars produce necrotic lesions on tubers. Leaf symptoms are particularly severe when PVY occurs as mixed infections with PVX.

Source of infection and spread

PVY can infect a wide range of hosts, especially solanaceous hosts. Susceptible crop plants in addition to potato include tomato, capsicum, chilli, tobacco and many weeds, including *Physalis* spp. (gooseberries, groundcherries), nightshades (*Solanum* spp.) and apple of Peru (*Nicandra physalodes*). Although the virus can spread by mechanical contact, the most important means of field spread is via aphids. Transmission can occur after feeding times of only a few seconds and over 25 species are recorded as vectors. The virus is also transmitted on infected tubers.

Importance

Internationally, PVY is one of the most serious viruses in potato, although it is uncommon in Australian potato crops.

Management

- Use certified seed potatoes.

- Insecticides may provide some control of virus spread within the crop, but they usually act too slowly to prevent spread by aphids entering the crop from outside sources of infection.

- Avoid growing young crops near older, infected potato crops and other virus hosts.

- Avoid mechanical damage to plants during crop maintenance.

- Sanitise equipment, especially during seed piece preparation.

■ *POTATO VIRUS S*

Cause
Potato virus S (Carlavirus, PVS).

Symptoms
Although frequently symptomless, the virus sometimes causes vein deepening, leaf rugosity, bronzing or necrotic spots or a more open plant habit.

Source of infection and spread
Spread by a number of aphid species through very short feeding times, and in infected tubers. The virus also spreads very easily by mechanical means, including plant-to-plant contact, on hands and equipment and during cutting of seed pieces.

Importance
PVS is the most frequently detected virus in potatoes worldwide, a situation reflected in Australia. Losses are usually in the order of 10–20% but can be exacerbated in the presence of other viruses.

Management
- Use certified seed potatoes.

- Insecticides may provide some control of virus spread within the crop, but they usually act too slowly to prevent spread by aphids entering the crop from outside sources of infection.

- Avoid growing young crops near older infected potato crops and other virus hosts.

- Avoid mechanical damage to plants during crop maintenance.

- Sanitise equipment, especially during seed piece preparation.

■ TOMATO SPOTTED WILT VIRUS

Cause
Tomato spotted wilt virus (Tospovirus, TSWV).

Symptoms
Necrotic spots or ringspots develop on leaves, tip necrosis, and necrotic streaks appear on stems. On susceptible varieties, tubers develop necrotic spots or rings and severe internal necrosis. Less susceptible varieties develop only occasional, internal spots and flecks.

The virus can be transmitted in tubers with the percentage of tuber infection varying considerably between varieties. It is unusual for all tubers from an infected plant to be infected.

Importance
Sporadic outbreaks of TSWV can cause considerable yield losses in potatoes in most production areas in Australia.

Refer to the chapter on Common diseases of vegetable crops.

Fig 19.43 *Tomato spotted wilt virus* showing necrotic ring symptoms on a leaf.

Fig 19.42 *Tomato spotted wilt virus* on potato leaves.

Fig 19.44 Severe, internal symptoms of *Tomato spotted wilt virus*.

POTATO SPINDLE TUBER VIROID – BIOSECURITY THREAT

Cause
Potato spindle tuber viroid (PSTVd).

Symptoms
Mild strains cause few if any symptoms on the foliage, especially in the earlier generations. More severe strains can cause plant stunting, upright growth habit and, when viewed from above, clockwise phyllotaxy. Leaves are often a darker green than normal, rugose and distorted. Tubers are elongated (or spindle-shaped) and may have surface cracks and very prominent eyes.

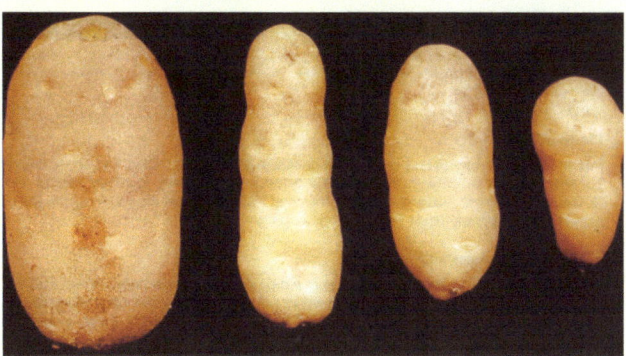

Fig 19.45 *Potato spindle tuber viroid* on tubers.

Source of infection and spread
PSTVd is readily transmitted by contact and can spread with plant-to-plant contact, and through mechanical injury via machinery or cutting implements. It is also transmitted through true seed of potato and tomato and through potato seed tubers. In mixed infections with *Potato leafroll virus*, PSTVd can also be simultaneously transmitted by aphids.

Importance
PSTVd is an important pathogen of potato and tomato. The viroid was detected on Australian potato breeding and germplasm stocks in NSW, Victoria and South Australia in the 1980s but was subsequently eradicated. More recently, it has been detected in tomato in NSW and Western Australia, and in a native *Solanum* species in the Northern Territory. It is a quarantine target disease in Australia.

What to do if you suspect *Potato spindle tuber viroid*
This pathogen is a biosecurity risk to Australia. Any suspected affected plants should be reported to the nearest Department of Primary Industries or the Plant Health Australia hotline (1800 084 881).

■ FURTHER INFORMATION

Banks E (2004). *Potato field guide: insects, diseases and defects*. Publication 823 Ontario Ministry of Agriculture, Food and Rural Affairs, Ontario, Canada.

Horne P, De Boer and Crawford D (2002). *Insects and diseases of Australian potato crops*. Melbourne University Press, Melbourne.

Johnson DA (Ed) (2007). *Potato health management* (2nd edn). APS Press, St Paul, Minnesota.

Stevenson WR, Loria R, Franc GD and Weingartner DP (Eds) (2001). *Compendium of potato diseases* (2nd edn). APS Press, St Paul, Minnesota.

Wale S, Platt HW and Cattlin N (2008). *Diseases, pests and disorders of potatoes – a colour handbook*. Manson Publishing, United Kingdom.

Rhubarb (*Rheum rhabarbarum*) is a herbaceous perennial in the family Polygonaceae. It is a cool season crop valued for its long, thick, red stems, which grow from the crown up to 75 cm in length. Stalks are used raw or cooked in pies, desserts and sauces. The crop is propagated each year by the division of two to three year old crowns.

Summer production occurs in cooler areas of southern Australia while winter production is concentrated in cooler areas of south-eastern Queensland.

Wet, cool weather favours downy mildew while virus infection has become an important industry problem.

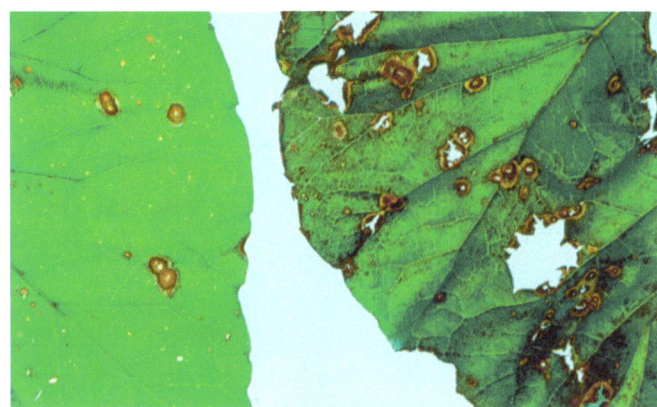

Fig 20.1 Leaf spot on rhubarb.

FUNGI

■ DOWNY MILDEW

Cause
The oomycete *Peronospora jaapiana*.

Symptoms
Light brown, angular lesions develop on the upper surfaces of leaves. Initially, major veins limit the lesions, then, as the disease progresses large areas of the leaves become brown. Eventually the brown areas tear, and sections of leaf fall out, giving leaves a ragged appearance. Severe defoliation can occur when weather conditions favour the disease. Under moist conditions, the downy growth of the fungus is visible on the undersides of affected leaves.

Source of infection and spread
Spores of the pathogen spread in wind during cool, wet weather. Downy mildew is unlikely to be a problem during dry weather.

Importance
The disease can be serious following periods of wet, cool weather.

Management
- Remove affected leaves.

- Apply the recommended fungicides as required.

■ LEAF SPOT

Cause
The fungus *Ascochyta rhei*.

Symptoms
Small, circular, brown spots develop on leaves. The spots enlarge, develop reddish-brown borders, and contain numerous small, black fruiting bodies of the fungus. Affected leaf tissue dies and falls out. Severe defoliation can occur if conditions favour the disease. The leafstalk also can develop small, oval, reddish-brown spots that increase to about 10 mm in length.

Source of infection and spread
The fungus survives in infected plants and crop residues. Spores spread in water splash and wet conditions favouring the disease.

Importance
Leaf spot is a relatively common disease in rhubarb.

Management
- Remove old leaves and stems from the crop.

- Apply the recommended fungicides as required.

■ RUST

Cause
The fungus *Puccinia rhei-undulati*.

Fig 20.2 Rust on rhubarb.

Fig 20.3 Symptoms of viral decline disease on rhubarb.

Symptoms

Small, raised pustules containing orange-red spores develop on the undersides of leaves and on the leafstalks. Small, brown spots occur on the upper surface of leaves.

Source of infection and spread

Fungal spores produced in the pustules spread in air movement and wind. Warm humid weather favours the disease.

Importance

Rust is a relatively common disease of rhubarb.

Management

- Remove affected leaves.

- Apply the recommended fungicides as required.

Fig 20.4 Leaf symptoms of viral decline disease on rhubarb.

VIRUSES

■ RHUBARB DECLINE DISEASE

Cause

Viruses, including *Rhubarb decline-associated closterovirus, Cucumber mosaic virus, Turnip mosaic virus, Tomato spotted wilt virus*.

Symptoms

The main symptoms of rhubarb decline disease are necrotic spotting and mottle symptoms on the leaves, and general stunting and reduced vigour of the plant. Symptoms become more prominent as the leaves mature, and more noticeable in cooler weather conditions. The exact cause of the disease is uncertain, as, frequently,

Fig 20.5 Chlorotic spot symptoms in rhubarb infected with *Tomato spotted wilt virus*.

Fig 20.6 Red ringspots in rhubarb infected with *Tomato spotted wilt virus*.

Fig 20.8 Rhubarb decline disease in a plant infected with *Rhubarb closterovirus* and showing chlorotic mottle and chlorotic and necrotic spots.

affected plants are infected with a mixture of viruses. *Rhubarb decline-associated closterovirus* and *Cucumber mosaic virus* are most commonly associated with these symptoms, though two unidentified viruses, as well as *Turnip mosaic virus* and *Tomato spotted wilt virus* are sometimes present.

Tomato spotted wilt virus alone causes symptoms of chlorotic spots of red rings on the leaves.

Source of infection and spread

Generally, rhubarb is propagated vegetatively as divided crowns, and all the viruses can be transmitted with this planting material. Most growers consider seedlings unsatisfactory, due to the plant-to-plant variation that can occur. It is not known whether any of the viruses are transmitted through rhubarb seed.

Aphids transmit *Rhubarb decline-associated closterovirus*, *Cucumber mosaic virus* and *Turnip mosaic virus*. Infection

of newly planted virus-free rhubarb can occur quickly if older, infected plantings are nearby. *Cucumber mosaic virus* has a very wide host range, including more than 1000 plant species, and harboured by a wide range of broad-leaved weeds. Turnip mosaic virus can also infect many plant species, especially in the Brassicaceae (e.g. cabbage, cauliflower, wild radish, stock).

Tomato spotted wilt virus also has a very wide host range, infecting mostly broad-leaved crop and weed species. It is transmitted by thrips.

Importance

Rhubarb decline is common in the eastern mainland states of Australia. It has not been reported from Tasmania, South Australia or Western Australia.

Management

- Use virus-free planting material. Selecting symptomless plants as a source of crowns can help reduce virus levels, but caution is needed, as this material is frequently already virus-infected without showing symptoms. Virus-free material can be produced through tissue culture, and if properly virus-tested, provides the most reliable source of planting material.

- Seedlings can be used as a source of planting material with minimal risk of virus contamination, but plant-to-plant variation may make this material less desirable, from a horticultural point of view.

- Remove older, disused plantings as soon as possible to remove this source of infection.

- Isolate new plantings to avoid re-infection from older, diseased crops.

Fig 20.7 Rhubarb decline disease symptoms including chlorotic mottle and chlorotic and necrotic spots. The plant is infected with *Rhubarb closterovirus* and *Turnip mosaic virus*.

Sweet corn (*Zea mays*), family Poaceae, is a summer vegetable crop, grown widely for both fresh market use and processing as kernels and frozen cobs. Standard sweet corn is based on high sugar levels from the sugary gene. Supersweet corn, based on the shrunken genes, provides higher sugar levels and delays the conversion of sugar to starch, which extends the shelf life of harvested cobs.

Fungal leaf diseases can cause serious damage in wet, humid conditions while *Johnson grass mosaic virus* can be serious in late season crops in Queensland.

BACTERIA

■ BACTERIAL LEAF STREAK

Cause
The bacterium *Xanthomonas campestris* pv. *zeae*.

Symptoms
Tan-brown elongated, angular lesions extend along the lengths of the leaves. Lesions are typically 2–3 mm wide and aligned parallel to the leaf veins but may coalesce to give leaves a blighted appearance. Following wet weather, white superficial exudate may be associated with the lesions.

Fig 21.1 Detail of bacterial leaf streak.

Fig 21.2 Bacterial leaf streak symptoms on leaves.

Source of infection and spread
The pathogen is presumed to survive in infected plant debris. Like many bacterial diseases, epidemics are favoured by warm, wet weather. It is not known if the pathogen is seed-borne.

Importance
Bacterial leaf streak is a minor disease of sweet corn in Australia. The disease has only been reported occasionally following warm, wet weather.

Management
• Specific control measures are not warranted.

FUNGI

■ BOIL SMUT (COMMON SMUT)

Cause
The fungus *Ustilago zeae*.

Symptoms
Boil smut infects maize, sweet corn and teosinte (*Zea mexicana*). The fungus attacks any actively growing, above-ground part of the plant to form swellings referred to as boils, blisters or galls. These are most common on cobs,

Fig 21.3 Boil smut symptoms on sweet corn stem.

stems (even at ground level) and tassels, but also develop on leaves. Young galls are pale green in colour and later develop a thin, white membrane that encloses a dark brown to black mass of spores. Mature galls, which are up to 200 mm in diameter, rupture to release the spores.

Source of infection and spread

Ruptured galls release spores (chlamydospores) spread in wind, with seed or stock food, and in soil adhering to clothes, vehicles, farm machinery and animals. Chlamydospores may survive in the soil for many years, and under favourable conditions germinate to produce aerial spores (sporidia) which spread in air currents and rain splash. The spores germinate, infect young tissues of the host and stimulate the host cells to proliferate and form galls.

Importance

Boil smut occurs in most production areas. Disease occurrence is usually sporadic and minor. However, high incidences can occasionally occur in sweet corn and maize crops.

Management

Specific control measures are not warranted. Most hybrids have at least a reasonable level of resistance to boil smut.

■ COMMON RUST

Cause

The fungus *Puccinia sorghi*.

Symptoms

Round to elongate, reddish-brown pustules develop on both the upper and lower surfaces of leaves. Pustules are up to 2 mm long and occur in scattered groups over the surface. Leaf spots arise when tissue dies around clusters of pustules.

Source of infection and spread

The fungus survives on diseased sweet corn or maize residues and volunteer plants. Large numbers of spores develop in the pustules and spread over long distances in

Fig 21.4 Boil smut on sweet corn ear.

Fig 21.5 Common rust.

the wind. Warm, humid weather favours leaf infection, disease development and spore production. The disease generally appears after tasselling and does not become severe until grain filling is complete. However, yield may be reduced if warm, wet weather occurs during the grain-filling period.

Importance
This is a common but usually minor disease. Late-planted crops are more likely to develop serious disease during the grain-filling period. Some supersweet hybrids are highly susceptible.

Management
- Destroy volunteer maize and sweet corn plants before sowing.

- Early planting avoids high disease levels.

- Plant resistant hybrids late in the season, when the disease is likely to be serious.

■ FUSARIUM COB ROT

Cause
Several fungi of the genus *Fusarium* including *Fusarium verticillioides*, *F. proliferatum* and *F. subglutinans*.

Symptoms
The symptoms that develop are affected markedly by the variety, environmental conditions and the severity of the disease.

Typically, white fungal growth covers individual kernels scattered throughout the ear or, in very severe infections, the entire cob. Infected kernels may develop a 'starburst' symptom with white streaks radiating from the point of silk attachment at the cap of the kernel or from the base.

Source of infection and spread
The fungi survive in crop debris and may be seed-borne as well.

Fig 21.6 Fusarium cob rot.

Air-borne spores infect silks and the pathogen grows down the silks as they start to die, and colonise the kernels. The fungus may also systemically invade the stalk and invade the ears through the ear shank. Infection may also occur following cob injury or damage from insect feeding.

Fusarium cob rot is usually associated with high temperatures and moisture stress.

Kernel splitting and damage from insects, particularly *Helicoverpa*, favours infection.

Pericarp thickness and the extent of husk cover also affect the extent of damage.

Importance
Cob rots can cause significant damage in susceptible varieties maturing under high temperatures and suffering moisture stress.

Management
- Manage irrigation to ensure adequate moisture levels during silking through to harvest.

- Avoid over-irrigation of sweet corn towards harvest, as this favours kernel splitting and subsequent Fusarium infection.

- Maintain good insect pest management to minimise cob damage.

- Avoid harvesting delays as this encourages kernel splitting and disease development.

■ JAVA DOWNY MILDEW

Cause
The oomycete *Peronosclerospora maydis*.

Fig 21.7 Java downy mildew on a sweet corn leaf.

Symptoms

Pale chlorotic or yellow striping, which resembles virus infection, develops on leaves. Leaf stripes have distinct borders between healthy and infected areas. On young infected plants, the base of the leaf is often chlorotic when it emerges, with each successive leaf having larger areas of affected tissue. White sporulation of the fungus can often occur on the chlorotic areas of young plants, but not on older plants.

Severe distortion of mature plants frequently occurs. Symptoms include multiple cobbing, deformed or leaf-like tassels and cobs, and elongated or shortened stems. Root lodging may increase becaue of the disease.

Source of infection and spread

The pathogen is probably endemic in northern Australia on plume sorghum (*Sorghum plumosum*). Maize plants are infected systemically soon after emergence, and become resistant to infection with age. Infection is through conidia, produced on plume sorghum growing near maize and sweet corn crops. Conidia develop at night and, as the conidia are short-lived, spread is restricted to within a short distance of the source.

Importance

The disease is restricted to the drier areas of north Queensland where plume sorghum is common. Java downy mildew can cause severe losses in sweet corn and maize.

Management

- Treat seed with a recommended fungicide before planting.

■ TURCICUM LEAF BLIGHT

Cause

The fungus *Exserohilum turcicum*.

Fig 21.8 Turcicum leaf blight on sweet corn (detail).

Fig 21.9 Turcicum leaf blight.

Symptoms

Symptoms generally develop first on the lower leaves. Long, spindle-shaped, greyish-green, water-soaked spots develop on leaves. The spots or lesions become grey and often exceed 100 mm in length. When weather conditions favour the disease, large areas of leaves are blighted. In moist weather, dark masses of spores cover the lesions.

Source of infection and spread

The fungus survives on maize and sweet corn crop residues and volunteer plants. Spores spread in wind and rain. Warm, wet weather favours the disease.

Importance

Leaf blight is a common disease in sweet corn. Severe blight before or during tasselling reduces cob fill and yield.

Management

- Plant resistant hybrids.

- Destroy volunteer plants and crop residues before sowing.

- Early planting usually avoids high levels of disease.

■ POLYSORA (TROPICAL) RUST

Cause

The fungus *Puccinia polysora*.

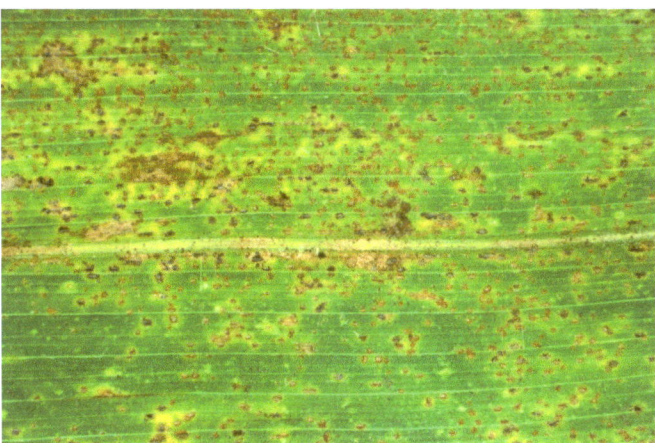

Fig 21.10 Polysora rust on sweet corn.

Fig 21.11 Maize stripe disease.

Symptoms

Small, round, reddish-brown pustules are distributed relatively uniformly over the upper surface of leaves. Severely affected leaves die prematurely. Pustules produced on midveins and leaf sheaths are larger than those on the leaves and are irregular in shape.

Source of infection and spread

The fungus survives on volunteer maize and sweet corn plants. The spores spread in wind and can carry over long distances. Warm, wet weather favours the disease.

Importance

In Australia, tropical rust has been recorded only in north Queensland. The disease can defoliate susceptible sweet corn hybrids.

Management

• Plant resistant hybrids in areas where the disease can occur.

VIRUSES AND VIRUS-LIKE DISEASES

■ MAIZE STRIPE DISEASE

Cause
Maize stripe virus.

Symptoms

Light specks occur on young leaves; these elongate, broaden and fuse in subsequent growth to form light yellow stripes parallel to the midrib. The stripes may extend into broad, white or yellow bands. Affected plants often bend at the apex, and those infected at an early growth stage are usually stunted.

Source of infection and spread

The virus persists from season to season on volunteer maize plants, sorghum and wild sorghums. It is spread by the maize planthopper (*Perigrinus maidis*).

Importance

Maize stripe is relatively common in sweet corn crops in coastal areas but is rarely serious.

Management

Management is not warranted. Resistant hybrids are not available.

■ MOSAIC

Cause

Johnson grass mosaic virus (Potyvirus).

Fig 21.12 Mosaic caused by *Johnson grass mosaic virus*.

Fig 21.13 Symptoms of *Johnson grass mosaic virus* on sweet corn.

Symptoms

Mosaic and ringspot patterns develop on leaves. Mosaic symptoms consist of light green and dark green patches, usually in the form of broken lines between the veins. Very susceptible hybrids show extensive leaf yellowing. Early infection often results in severe stunting and considerable yield reduction.

Fig 21.14 Ringspotting symptom of *Johnson grass mosaic virus* on sweet corn.

Source of infection and spread

Johnson grass (*Sorghum halepense*) is the main perennial host of the virus. It also survives in stand-over or ratooned forage and grain sorghum crops. The virus is spread by aphids. As these insects can transmit the virus during very short feeding periods, insecticides are of little value in disease control.

Importance

Mosaic is a serious disease in susceptible hybrids, particularly in crops planted late in the season.

Management

- Plant resistant hybrids.
- Spring or early season plantings usually avoid high disease levels.

■ WALLABY EAR

Cause

Once thought to be due to virus infection, it is now known that the symptoms result from a toxin injected by leafhoppers while feeding.

Symptoms

Leaves of affected plants are dark green and are short and stiff, often with enlarged, protruding leaf veins. Affected plants may be severely stunted, particularly if the leafhoppers are present during the early stages of growth.

Source of infection and spread

The small, pale-coloured leafhopper (*Cicadulina bimaculata*) breeds on various grasses and is common on

Fig 21.15 Wallaby ear disease on sweet corn.

Fig 22.4 Foliage scab on leaves.

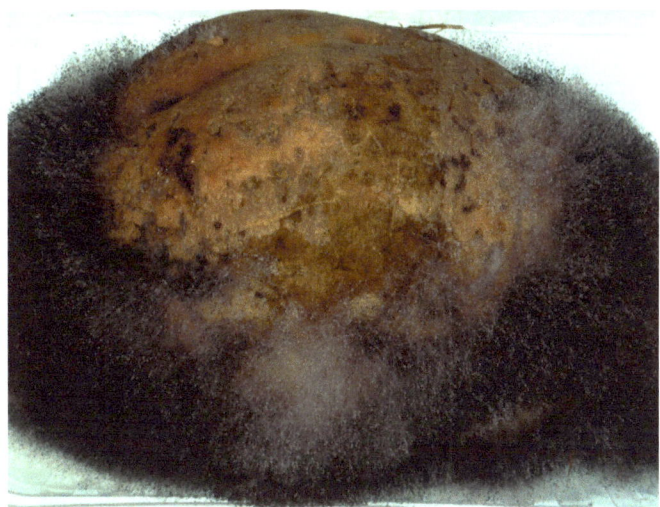
Fig 22.5 Rhizopus storage rot in a sweetpotato root.

Source of infection and spread

The fungus survives in the regrowth and debris from infected crops. The main source of inoculum is spores produced from cushions of fungal tissue that rupture the surface of lesions. Spores disperse in rain splash. Planting material collected from infected regrowth introduces the pathogen into new crops. The disease is favoured by wet weather over a wide temperature range (13–26°C).

Importance

Scab is the most severe foliar disease of sweetpotato.

The disease occurs throughout Asia, the Pacific region and has been recorded in northern Queensland. It is severe when infected regrowth of previous crops is used as planting material. The importance of this disease in Australia is low due to selection of varieties with high levels of tolerance.

Management

- Destroy regrowth of infected crops.

- Use healthy planting material established from healthy roots or stems, produced in a nursery.

- Plant resistant varieties when available.

■ STORAGE ROT

Cause

The fungi *Rhizopus* spp.

Symptoms

The rot may begin at one end of the root or on the shoulder, causing a sunken lesion that tends to encircle the root. A soft, watery rot develops under humid conditions. The root is covered by a mass of greyish-white fungal growth in which numerous small, black, stalked fruiting bodies (sporangia) develop, giving the appearance of whiskers.

Source of infection and spread

Rhizopus species are ubiquitous in soil and in the atmosphere. Spores contaminate harvested roots and enter through wounds. Roots and cuttings used as planting material also can be invaded. Relative humidity of 75–85% favours the development of rot; a drier atmosphere inhibits mould development.

Importance

Storage rot is a common and sometimes serious post-harvest disease of sweetpotato in most countries, but of limited importance in Australia, where storage is minimal because the crop is harvested and marketed all year round.

Management

- Dip or spray planting material with a registered fungicide before planting.

- Handle harvested roots carefully to minimise injury.

NEMATODES

■ ROOT-KNOT NEMATODE

Cause

The nematode *Meloidogyne* spp.

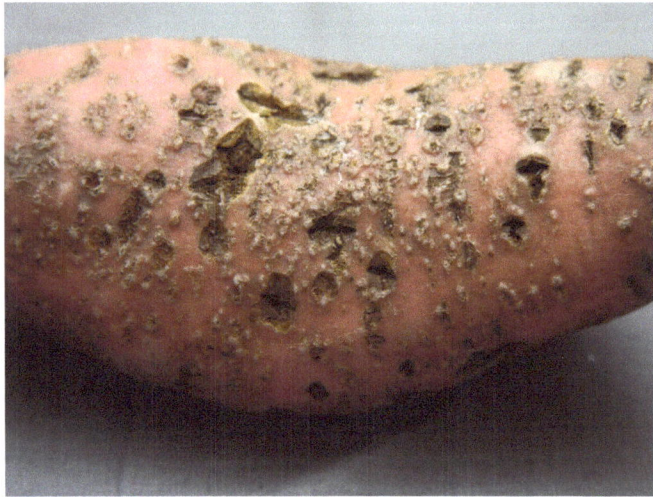

Fig 22.6 Symptoms of root rot nematode in sweetpotato.

Refer to the chapter on Common diseases of vegetable crops.

VIRUSES

■ FEATHERY MOTTLE

Cause
Sweet potato feathery mottle virus (Potyvirus).

Symptoms
Symptom development is influenced by the variety, rate of growth, plant age and weather conditions. Symptoms include a mild mottle on young leaves, yellow (chlorotic) spotting often surrounded by a purple margin or irregular chlorotic patterns along the midrib. Symptoms may be

Fig 22.7 *Sweet potato feathery mottle virus* on leaves.

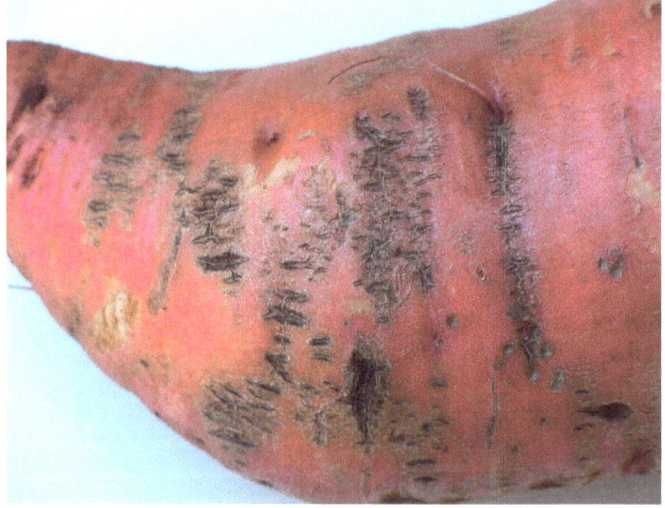

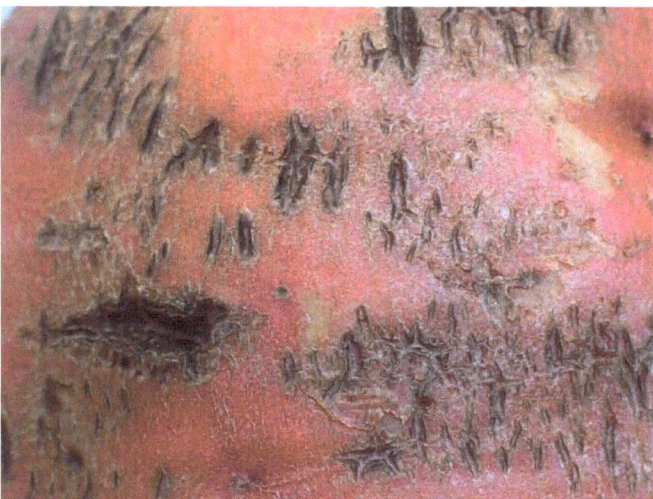

Fig 22.8 Russett crack symptoms on storage roots infected with *Sweet potato feathery mottle virus*.

more prevalent on older leaves. Plants of many varieties develop only transient, mild symptoms. Reduced plant vigour in the early establishment phase of the crop is common but not recognised until infected material is grown alongside PT material. Storage roots elongate and become thinner in successive generations where infected plant material is used. Storage roots can display russet cracking and pale flesh colour in orange-fleshed varieties.

Source of infection and spread
The virus is spread by aphids in a non-persistent manner and by using infected planting material. The host range of the virus is restricted to sweetpotato and other *Ipomoea* species.

Importance
Sweet potato feathery mottle virus is the most common and widespread virus infecting sweetpotato, occurring in

almost all countries where the crop is grown. The virus can reach high levels of infection in crops propagated from infected material or grown near infected crops, allowing aphid transmission from crop to crop.

The virus causes reduced yields and a reduction in the quality of harvested roots.

Management

- Use only good-quality planting material, preferably from tested, virus-free material.

- Locate planting material nurseries remote from sweetpotato crops to reduce the risk of infection from diseased crops.

- Discard plants with probable symptoms of virus disease.

■ FURTHER INFORMATION

Clark CA and Moyer JW (1998). *Compendium of sweetpotato diseases.* APS Press, St Paul, Minnesota.

The cultivated tomato *Solanum lycopersicum* is a member of the Solanaceae family. This family includes several other important cultivated species including potato, capsicum, eggplant and tobacco, which share many of the same diseases as tomato. Tomatoes are grown in all Australian states, with Victoria, Queensland and New South Wales being the major producers. Most fresh market production occurs in Queensland, while Victoria and New South Wales produce large quantities for processing. New Zealand, Singapore and Hong Kong are important export markets for Australian fresh market tomatoes.

A large number of diseases affect tomatoes and effective crop protection is a key aspect of profitable tomato production.

BACTERIA AND PHYTOPLASMAS

■ BACTERIAL CANKER

Cause
The bacterium *Clavibacter michiganensis* subsp. *michiganensis*.

Symptoms
The main symptom of bacterial canker is plant wilt, although the symptoms of the disease are variable and may differ between field-grown and greenhouse crops.

Fig 23.2 Bacterial canker showing brown discolouration of the vascular tissues.

Field crops: Symptoms usually develop first on the lower leaves with leaves turning downwards; death of leaflet margins to give a leaflet 'firing' symptom that may easily be confused with fertiliser burn, and which may begin on one side of the leaflet only. The wilted leaflets gradually turn yellow and die.

Light brown cankers may develop on the stems, although these may not be obvious in all cases. When a diseased stem is cut, a brown discolouration of the water-conducting tissues is seen. The discolouration may initially be quite mild, but with time the discolouration extends into the

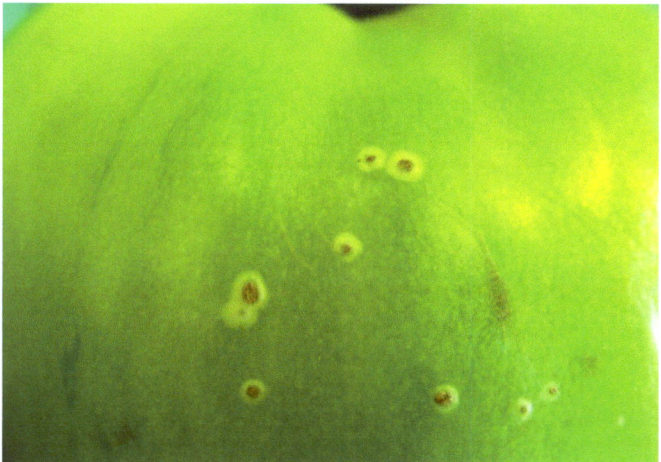

Fig 23.1 Bacterial canker on tomato fruit showing the 'birds eye' spot symptom.

Fig 23.3 Leaflet 'firing' in field grown tomatoes affected by bacterial canker.

Fig 23.4 Wilting tomato leaves with bacterial canker, showing 'one-sided' symptoms.

leafstalks, the stem pith tissues become brown and collapse, and the stem develops a mealy appearance. In advanced infection, cavities develop in the stem and the stems become desiccated.

Fruit spots was once a common diagnostic symptom of bacterial canker but is now uncommon in field or protected cropping. When fruit spots occur, they are circular, up to 3 mm wide, with a central, raised, brown area surrounded by a pale halo. Owing to their appearance, they are known as bird's eye-spots. As the spots age, the central areas crack, giving a ragged appearance.

Greenhouse crops: The leaflet symptoms may differ from those seen in field crops. Leaflet wilting in the greenhouse does not necessarily begin at the base of the plant. Instead, sudden wilting, which is often unilateral, affects the leaflets at any location on the plant. Leaflet wilting develops as

Fig 23.5 Interveinal necrosis is a common symptom of bacterial canker in greenhouse-grown tomatoes.

Fig 23.6 Glasshouse symptoms of bacterial canker. Sections of an infected trellis tomato crop have died.

pale, greasy, interveinal areas that dry to form white to beige patches on the leaflets. Leaflet yellowing does not always accompany wilting in the greenhouse. Very rarely, small cankerous spots may develop on the leaflets, stems and fruit.

Source of infection and spread

The bacterium is commonly introduced in seed. Other sources of infection include infected crop residues, weed hosts, volunteer tomato plants and contaminated stakes and trellis wires. The bacterium is highly infectious and the disease has a long latent period, with symptoms often not developing until after transplant from the nursery. The most common source of spread from infested to healthy plants is via sap transfer during pruning and leaf removal. Rain-splash, overhead irrigation, nutrient solutions in soil-less crops, transplanting and trellising operations and contaminated tools and machinery can also spread the bacterium. Like most bacterial diseases, canker is favoured by warm, wet and humid weather. Temperatures of 18–24°C with greater than 80% humidity favours disease development.

Importance

Bacterial canker is a very serious disease of both field and greenhouse tomatoes in Australia, particularly in cooler growing areas.

Management

- At present, there are no commercially acceptable resistant varieties available.

- Use disease-free seed. Seed suspected of being infected should not be planted. Treating seed in hydrochloric acid or hot water will reduce the bacterial load but will not eliminate the pathogen from the seed.

- Raise seedlings in sterilised soil or soil-less potting mix in trays or pots.

- Avoid planting areas containing undecomposed tomato residues. Allow at least one year before replanting an area where the disease has occurred.

- Check for affected plants during trellising and pruning. Carefully remove the young affected plants if numbers are small.

- Disinfect stakes and trellis wires before reuse.

- Disinfect pruning equipment and hands at regular intervals (e.g. at the start of each line).

- Remove and dispose of pruned debris from the crop.

- Apply foliar copper sprays regularly after transplant, particularly after rain or high humidity or in areas where the disease has previously occurred.

■ BACTERIAL SPECK

Cause

The bacterium *Pseudomonas syringae* pv. *tomato*.

Fig 23.7 Bacterial speck on fruit. Inset: symptoms on the fruit stem and calyx.

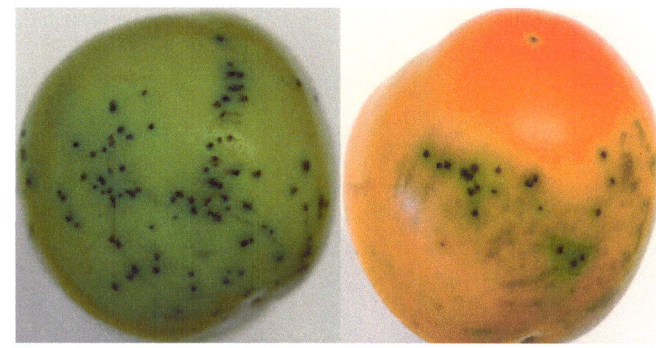

Fig 23.8 Speck symptoms on green and ripe fruit.

Symptoms

Small, black, greasy spots develop on leaves and stems and are similar to, but generally smaller than, those caused by bacterial spot. Lesions on leaves are angular to round and initially lack a halo. A halo may develop in time however, and when spotting is severe, the entire leaf yellows. On the stem, spots are more elongated. Raised, black specks, rarely more than 1 mm wide, occur on fruit. These may coalesce to form large spots similar to those of bacterial spot.

Source of infection and spread

The bacterium is seed-borne. Once established, it spreads during wet, windy weather and by overhead irrigation water. High humidity (greater than 80% relative humidity), free moisture and cool temperatures (18–24°C) favour the disease. Although the bacterium is not soil-borne, it may survive in the soil on undecomposed plant residues for as long as 30 weeks. Alternative weed hosts may also harbour this pathogen.

Importance

Bacterial speck is generally a minor disease of tomato but outbreaks can be severe and difficult to control in wet weather.

Fig 23.9 Bacterial speck on tomato leaves.

Fig 23.10 Advanced symptoms of bacterial speck.

Management

- Spray with the recommended chemicals. Mixes of copper and mancozeb are more effective than foliar copper sprays alone.

- Use the recommended seed treatment, e.g. hydrochloric acid or hot water.

- Destroy old crops promptly and remove tomato volunteers.

- Regularly sanitise tools, pruning equipment and worker's hands with a disinfectant.

- Produce disease-free seedlings in nursery areas away from tomato crops.

■ BACTERIAL SPOT

Cause

The bacterium *Xanthomonas campestris* pv. *vesicatoria*. Numerous races of this pathogen have been identified,

Fig 23.12 Bacterial spot on tomato stem.

including races pathogenic only to tomato, races pathogenic specifically to capsicum and races pathogenic to both hosts.

Symptoms

Small circular to irregular, greasy spots develop on leaves and stems, becoming dark tan to black. As the disease

Fig 23.11 Bacterial spot on tomato leaves.

Fig 23.13 Severe bacterial spot on young fruit.

progresses, the lesions may coalesce to give large necrotic areas on leaves. Leaves turn yellow and fall.

The initial symptoms on fruit are small, raised, circular, black spots with a distinct water-soaked margin. As the fruit enlarges, spots increase in size and become slightly sunken with a central scab. Only the outer skin is affected. Mature fruit are resistant to infection.

Source of infection and spread

The bacterium is commonly introduced in seed. Once established, it spreads during wet, windy weather and in overhead irrigation water, entering plants through stomata, wounds and insect punctures. Warm temperatures (24–30°C) and humid weather favours the disease. Although the bacterium is not soil-borne, it may survive in the soil on undecomposed plant residues. The pathogen may also carry over on volunteer tomato plants and on weeds such as black nightshade (*Solanum nigrum*).

Importance

Bacterial spot is a sporadic disease but it can be serious and difficult to control in wet weather.

Management

- Spray with the recommended chemicals. Mixes of copper and mancozeb are more effective than foliar copper sprays alone.

- Use the recommended seed treatment.

- Destroy old crops promptly and remove tomato volunteers.

- Regularly sanitise tools, pruning equipment and worker's hands with a disinfectant.

- Produce disease-free seedlings in nursery areas away from tomato crops. Spray with the recommended chemicals.

■ BACTERIAL WILT

Cause

The bacterium *Ralstonia solanacearum*.

Symptoms

Bacterial wilt first appears as a drooping of one or more of the youngest leaves. A rapid 'green' wilting of the foliage then develops, which is particularly noticeable during the warmest part of the day. Stunting may precede the wilting, and leaflets and leafstalks may curl downwards. Where the disease has developed slowly, many adventitious roots are produced along the stem above ground level.

Fig 23.14 Internal stem symptoms of bacterial wilt.

In the early stages of the disease, if the stem is cut across near ground level, a yellow-light brown discolouration of the water-conducting tissues just beneath the bark can be seen. This discolouration becomes darker as the disease progresses and in advanced cases, the stem cortex and pith tissues also become brown. Extensive, dark brown water-soaked lesions may also develop on the external stem, particularly at the base of the plant. For plants at an advanced stage of infection, a milky exudate is apparent after several minutes if the end of a cut stem is placed in a glass of water.

Below-ground symptoms are variable. Initially, root systems may appear healthy with the only evidence of infection being bacterial invasion through the root tips. Only one or a few roots may show a brown rot, but as the disease progresses, the entire root system may become discoloured and fine feeder-roots will be lost.

Fig 23.15 Symptoms of bacterial wilt are rapid foliage wilting, usually without yellowing.

Source of infection and spread

The bacterium affects a very wide range of hosts among cultivated plants and weeds. Most economically important hosts are in the potato family (*Solanaceae*) and include potato, tomato, eggplant, tobacco and capsicum. The pathogen infects the roots of alternative weed hosts and survives in the soil for extended periods in the absence of host plants.

Ralstonia solanacearum enters roots through wounds made during transplanting and cultivation, or by nematodes or soil insects. It rapidly multiplies in the water-conducting tissues of plants, filling the tissues with bacterial cells and slime thereby preventing water uptake and transport, and causing the plant to wilt. Disease development occurs rapidly at high temperatures (30–35°C).

The bacterium is spread through fields in running water and via soil movement, and root contact may also cause plant-to-plant spread. Long-distance dispersal can occur through the movement of infected plants, especially seedlings.

Importance

Bacterial wilt may be a serious disease in the field during warm weather. It also is potentially severe in greenhouse production.

Management

- Avoid planting into known infested areas in summer.

- Use fumigated soil or soil-less mix in seedling nurseries. Do not apply animal manure or excessive amounts of nitrogenous fertilisers.

- Resistant cultivars are available and are particularly useful as rootstock material. Resistance may break down during periods of high soil temperatures.

- Rotations are of limited value because of the pathogen's wide host range and its ability to survive in soil.

■ TOMATO PITH NECROSIS

Cause

The bacterium *Pseudomonas corrugata* is the most frequently reported causal agent worldwide, although the same symptoms have also been attributed to *Pseudomonas viridiflava*, *Pseudomonas fluorescens* biotype I and *Erwinia carotovora*.

Symptoms

Symptoms of tomato pith necrosis resemble those of bacterial canker. The disease affects older plants and symptoms usually do not show until fruit begin to develop.

Fig 23.16 A cut stem section showing bacterial exudate in water.

Initial symptoms include a chlorosis of the young leaves. In severe infection, leaf chlorosis and wilting may occur together at the top of the plant and there may be a necrosis of the lower stem. Affected leaves may turn brown at their margins and curl up, and dark brown-black lesions may develop on the surfaces of lower stems. Internally, stems are darkly discoloured and the pith tissues contain cavities, which extend to produce hollow internal stems. Profuse development of adventitious roots is associated with the areas where the pith is affected. Occasionally, plants may die when the lower stem is affected, however although initially symptoms appear to be destructive, the disease does not progress and it may become undetectable as the crop grows.

Source of infection and spread

The epidemiology of tomato pith necrosis is not well understood, although it appears to be associated with low night temperatures, high nitrogen levels and high

Fig 23.17 Tomato pith necrosis in the field.

Fig 23.18 A cut tomato stem showing pith breakdown.

humidity. It occurs most frequently when the first fruit set is near the mature green stage.

Importance
Tomato pith necrosis is minor disease of tomatoes in Australia and it may be easily confused with bacterial canker. Plants with suspected tomato pith necrosis should therefore also always be tested for bacterial canker.

Management
- Avoid over-fertilising with high nitrogen fertilisers.
- Destroy old crops promptly and remove tomato volunteers.
- Regularly sanitise tools, pruning equipment and workers' hands with a disinfectant.

■ BIG BUD

Cause
Tomato big bud phytoplasma.

Symptoms
Symptoms in tomato may not appear for up to six weeks after infection. Apical stems become thickened, stiff and erect. Internodes are short-ended and flower buds are greatly enlarged with green petals. Leaves are small and distorted, with yellow and purple discolouration.

Fig 23.19 Field symptoms of tomato big bud disease.

Fig 23.20 Tomato big bud disease.

Source of infection and spread
Many cultivated plants and weeds are susceptible. The organism is spread by leafhoppers, particularly *Orosius argentatus,* which mainly breeds on weed hosts but migrates to tomato when weeds are drought-stressed. The disease is generally more common when dry weather causes the vegetation around a crop to dry off and leafhoppers move into the crop. There is very little spread of the disease within tomato crops.

Importance
Big bud is usually a minor disease in tomatoes.

Management
- Keep headlands free from weeds.
- Insecticides will provide control of leafhopper vectors.

■ SOFT ROT

Cause
The bacterium *Erwinia* spp.

Refer to the chapter on Common diseases of vegetable crops.

Fig 23.21 Bacterial soft rot in tomato.

FUNGI

■ ALTERNARIA ROT

Cause
The fungus *Alternaria alternata*.

Symptoms
Dark brown to mouldy black sunken spots develop on fruit and enlarge as the fruit mature. Spots generally appear on the shoulders of the fruit around the edges of the stem scar and around growth cracks or other small injuries to the skin (sunburn, blossom end rot, insect bites). On leaves, the fungus has been recovered only from dead tissues, most commonly around powdery mildew spots. Latent infections develop on fruit that have been cool-stored for prolonged periods. Symptoms are similar to Stemphylium fruit rot.

Source of infection and spread
The fungus survives on crop debris and spreads as windborne spores. The disease can be serious in varieties with fruit prone to growth cracks, as cracks allow entry of the fungus. Cool, wet periods favour the disease. Damage is particularly severe following heavy rain. Free water present on the fruit or between the fruits for several hours encourages development.

Importance
Alternaria rot is a post-harvest disease that is occasionally serious following wet weather.

Management
- Store the fruit at correct temperatures.
- Apply the recommended fungicides after harvest.

■ ANTHRACNOSE

Cause
Several species of the fungus *Colletotrichum*: *C. coccodes*, *C. gloeosporioides*, and *C. dermatium*. *Colletotrichum coccodes* is the species most frequently associated with fruit symptoms.

Symptoms
This disease primarily affects the fruit. Young, green fruit may be infected but symptoms do not develop until the fruit begin to ripen. On ripe fruit, small, circular, slightly sunken, water-soaked spots develop. The production of masses of pink-orange spores cause the spots to turn pink, and which then become dark as small, black fruiting bodies of the fungus develop. The spots may be up to 10 mm wide.

Source of infection and spread
The fungus survives on crop residues and weed hosts and in the soil. Spores splash on to fruit during wet, windy weather.

Fig 23.22 Alternaria rot in tomato fruit.

Fig 23.23 Anthracnose in tomato fruit.

Susceptibility to infection increases as the fruit ripen. The fungus can establish latent infections in immature fruit, resuming development as the fruit ripens. Temperatures 20–24°C are optimal for disease development.

Importance
Anthracnose is usually a minor disease of tomato.

Management
• Spray with the recommended fungicides.

• Destroy crop residues promptly.

■ FUSARIUM FOOT ROT

Cause
The fungus *Fusarium solani*.

Symptoms
Affected plants show delayed development, yellowing and death of leaves, become stunted and eventually wilt. A dry, brown rot of the crown and roots is evident. The rot is confined to the crown and root tissues, which is an important distinction from Fusarium wilt. Damping-off may occur in seedlings.

Source of infection and spread
The fungus is soil-borne and enters the feeder roots and moves slowly into the taproot and lateral roots. The pathogen also survives in tomato crop debris.

Importance
The disease occurs in the Bowen district in Queensland where it may be serious.

Fig 23.24 Fusarium foot rot showing leaf chlorosis and stunted plants.

Management
• Rotate tomatoes with non-host crops, e.g. legumes such as soybean and lablab.

■ FUSARIUM CROWN AND ROOT ROT

Cause
The fungus *Fusarium oxysporum* f.sp. *radicis-lycopersici*.

Symptoms
The initial symptom is chlorosis of the lower leaves, which progresses to a leaf necrosis. As the disease progresses, successively younger leaves become chlorotic and then necrotic. Infected plants can be stunted and unthrifty or they may decline more rapidly and suffer complete collapse. A dark brown discolouration develops in the root tissues and extends into the vascular tissues of the crown. This discolouration remains confined to the roots and crown however, and does not extend up the stem.

Source of infection and spread
Like Fusarium wilt pathogens, *Fusarium oxysporum* f.sp. *radicis-lycopersici* is a soil-borne fungus that can persist in

Fig 23.25 Fusarium crown and root rot in tomato. Note the dark discolouration in the plant crown.

soils for many years due to the production of resilient spores. The pathogen can also persist on tomato crop residues. In greenhouses, plants may become infected via air-borne microconidia. Optimum disease development takes place at temperatures between 20–22°C. Most commonly, the fungus enters the feeder roots and progresses into the tap and lateral roots, primarily as intercellular hyphae through the cortex and secondarily through the xylem.

Importance

The distribution of this pathogen in Australia is not well known. Generally, the disease is only sporadic, but when it does occur, it may be severe.

Management

- Some commercial varieties are purported to have resistance to this disease. The effectiveness of this resistance is against the disease in Australia is not well known.

- Take steps to avoid moving the fungus from infested to clean areas on vehicles and machinery.

- Decontaminate machinery by washing with the detergent-based degreaser Farmcleanse®.

■ FUSARIUM WILT

Cause

The fungus *Fusarium oxysporum* f.sp. *lycopersici*.

Symptoms

If young seedlings are affected the plants are generally stunted and show poor growth. More usually however,

Fig 23.27 Fusarium wilt in ground-type tomatoes.

symptoms are visible in older plants. Often, the first symptom is yellowing in the older leaves near the base of the plant, followed by general wilting of the foliage. Wilting may be particularly noticeable in warm weather. Sometimes only one branch of the plant shows definite symptoms. As the disease progresses, the whole plant will become chlorotic and wilt and the plant will eventually collapse and become desiccated.

Fig 23.26 Fusarium wilt in field-grown trellis tomatoes.

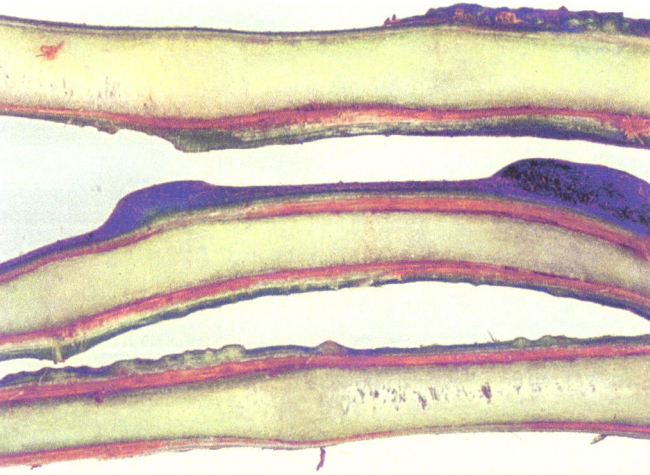
Fig 23.28 Red-brown discolouration of stem vascular tissues infected by *Fusarium oxysporum*.

If the bark is stripped from the plant just above ground level, a red-brown discolouration of the water-conducting tissues can be seen. This discolouration is often apparent a long way up the stem, which helps to distinguish this disease from Fusarium root rot, in which tissue discolouration is confined to the crown and roots. The pith remains healthy until the very late stages of the disease. Disease symptoms may be accentuated if the plant is bearing a heavy load of fruit or if it is stressed by some other factor. Affected plants will collapse and die prematurely.

Source of infection and spread

Fusarium oxysporum f.sp. *lycopersici* is soil-borne and remains in infested soils for many years. The fungus penetrates tomato roots and grows through the water-conducting system of the stem and leaves. It spreads in contaminated soil, infected transplants and in soil wash. Disease development is favoured by warm weather and high soil moisture. This pathogen is apparently host specific to tomato and three races are known. Races 1 and 3 occur in all districts of Queensland, whereas race 2 has been recorded only from the Bowen district in north Queensland.

Importance

There is now good resistance to all three races in both determinate and indeterminate tomato cultivars. Virtually all commercial varieties have resistance genes for races 1 and 2 and so outbreaks of disease caused by these two races are now extremely rare. Many of the commercial indeterminate types do not have resistance to race 3, and hence, outbreaks of race 3 of Fusarium wilt may still be serious if these varieties are grown.

Management

- Use the recommended varieties with resistance to races 1, 2 and 3.

- Take steps to avoid moving the fungus from infested to clean areas on vehicles and machinery.

- Decontaminate machinery by washing with the detergent-based degreaser Farmcleanse®.

■ GREY LEAF SPOT

Cause

The fungus *Stemphylium solani*.

Symptoms

Small (1–3 mm), dark brown to grey spots are scattered over the leaf surface. With age, the spots become lighter in

Fig 23.29 Grey leaf spot on tomato.

the centre and develop a pale-yellow halo. Severely affected leaves turn yellow.

Source of infection and spread

The fungus spreads as wind-borne spores. Disease development is favoured by warm, humid weather. The fungus survives on tomato crop residues, volunteer tomato plants and some weed hosts.

Importance

Grey leaf spot is now a minor disease because of the use of resistant varieties. Some cherry tomato varieties are susceptible.

Management

- Use resistant varieties.

- Apply the recommended fungicides, if necessary.

Fig 23.30 Grey leaf spot caused by *Stemyphillium*.

■ GREY MOULD AND GHOST SPOT

Cause
The fungus *Botrytis cinerea*.

Symptoms
Infections usually occur at a point of damage or where old blossoms fall onto the leaves, stem or fruit. A water-soaked, greyish area develops that is soon covered by a grey-brown, furry mould. Stems or branches may be girdled, causing the foliage to wilt. Flowers are often blighted.

Two types of fruit symptoms occur. One consists of pale, ring-like spots or haloes known as ghost spots on otherwise healthy fruit. These spots are the result of restricted development of the fungus following fruit infection. The other fruit symptom is a soft rot with growth of the grey-brown, furry mould on affected surfaces. Hard, black sclerotia often develop at the stem end of the fruit in the later stages of the disease.

Source of infection and spread
Botrytis cinerea has a very wide range of hosts. Spores are carried by wind and are often produced in enormous numbers on the furry mould on infected plants. The fungus can survive from season to season as hard-walled sclerotia in the soil or in crop residues. *Botrytis* is usually considered to be a weak parasite, although the fungus can penetrate plant tissues directly through the formation of appressoria if provided with a food source. Grey mould infection of foliage is usually associated with some form of injury or insect damage, with senescing leaves often invaded.

Cool weather with heavy dews and fog favours the disease. Long periods of high humidity are not required for infection; the humid conditions within the crop canopy often provide sufficient moisture for disease development.

Fig 23.32 Ghost spot symptoms of grey mould infection on green and ripe fruit.

Importance
Grey mould is a common and often serious disease of tomatoes, particularly in greenhouse cropping.

Management
- Minimise damage to the crop during cultural operations, especially trellising.

- Do not slash the tops of plants during periods when cool, moist conditions are expected.

- Avoid having sequential plantings close together.

- Spray with the recommended fungicides, rotating between fungicide groups and tank-mixing with protectant fungicides to minimise the chance of fungicide resistance developing in populations of the pathogen.

Fig 23.31 Grey mould on tomato fruit.

■ LATE BLIGHT

Cause
The oomycete *Phytophthora infestans*.

Symptoms
The disease first appears as pale green water-soaked patches on the leaves. The patches expand into large grey to dark brown leaf lesions. In moist weather, affected leaves develop a white, downy growth on the underside. With a return to fine conditions, affected areas become dry and papery. Stems and petioles may also develop similar symptoms. The pathogen may rapidly colonise tomato foliage and cause entire plants to turn brown, wither and die.

Fruit symptoms are first visible at the stem scar as small, greenish-brown spots that rapidly develop to produce large, mottled brown areas with an indefinite margin. The rot is generally firm, but the surface often becomes rough as the decay advances. If affected fruit are stored under humid conditions, a white, downy growth may develop on affected areas, particularly over the stem scar. Fruit not showing symptoms of the disease at harvest may develop extensive breakdown during transit and storage.

Source of infection and spread
Tomato and potato are the main hosts, although other members of the Solanaceae may also harbour the pathogen. The pathogen may carry over in volunteer host plants, and potato cull piles. When high humidity and cool temperatures prevail, the white, downy growth that develops on affected foliage produces large numbers of spores that spread in the wind to susceptible crops. Cool, wet weather (optimum temperature range 18–22°C) is essential for infection and disease development.

Importance
Late blight is usually a minor disease in Queensland. Dry weather usually occurs during winter, which is the main production period, thus preventing serious outbreaks of the disease that is favoured by prolonged periods of cool, humid weather.

■ LEAF MOULD

Cause
The fungus *Fulvia fulva*.

Symptoms
Early symptoms are pale green-yellow spots that develop first on the upper leaf surfaces of the older leaves. As the disease progresses, the spots become distinctly yellow and a green-grey, velvety fungal growth appears on the underside of each spot. Affected areas die prematurely, leaving the plant with a ragged appearance.

Source of infection and spread
The velvety mould produced on the underside of affected leaves contains large numbers of spores that readily spread

Fig 23.33 Late blight in tomato stem.

Fig 23.34 Leaf mould in tomato.

in wind. Warm, humid weather favours infection. The disease does not develop at humidity less than 85%. The fungus survives as a saprophyte on crop residues or as spores or sclerotia in soil. The spores readily spread in rain or wind and can be transmitted via contaminated tools, on workers' clothing and possibly by insects.

Importance
Leaf mould can be a serious disease during periods of warm, humid weather. It is of particular concern in protected cropping systems.

Management
- Spray with the recommended fungicides.
- Promptly remove and/or destroy crop residues

■ PENICILLIUM ROT

Cause
The fungi *Penicillium* spp.

Symptoms
Penicillium rot usually affects only poor-quality and over-ripe fruit. Dark olive-green or blue spores appear over wounds. Other fungi are often associated with the *Penicillium* spp. The rot is slow to develop.

Source of infection and spread
The main source of the fungi is diseased fruit in and around packing sheds. Infection occurs by spores entering through skin injuries, although these may be small and scarcely visible. The disease is more prevalent during wet weather, when fruit is easily damaged and conditions are ideal for the spread of the fungi.

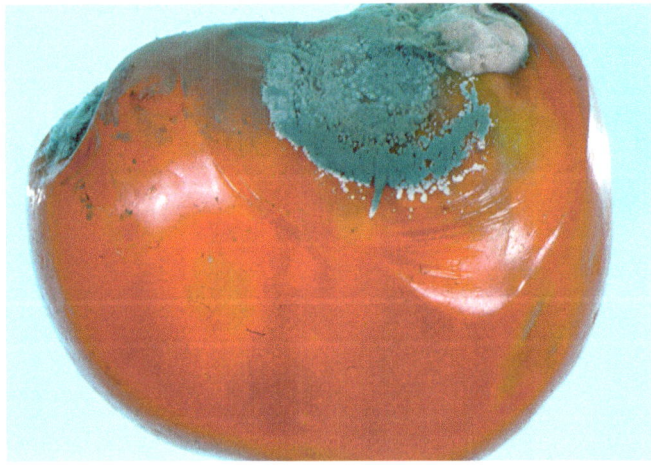

Fig 23.35 Penicillium rot in tomato fruit.

Importance
Penicillium rot is usually a minor disease of harvested fruit.

Management
- Handle fruit carefully to avoid skin damage.
- Discard fruit with serious growth cracks and other injuries.
- Remove reject fruit from the vicinity of the packing shed.
- Treat fruit with the recommended post-harvest fungicide.

■ PHOMA ROT

Cause
The fungus *Phoma destructiva*.

Symptoms
Dark, sunken spots develop on cracks and other damaged areas of both green and ripe fruit. Close examination of the spots reveals a pimpled appearance in the centre. The diseased tissue remains quite firm unless invaded by secondary organisms.

Source of infection and spread
The fungus survives in crop residues. During periods of wet weather, masses of spores are released from fruiting bodies of the fungus. The spores spread in splashing rain or irrigation water. The fungus can also be carried on tomato seed. Fruit infection occurs through wounds and develops rapidly in ripe fruit.

Importance
Phoma rot rarely occurs and it is usually only a problem during very wet seasons.

Management
- Handle fruit carefully during harvesting and packing to avoid injury.
- Spray with recommended fungicides.

Fig 23.36 Phoma rot in tomato fruit.

■ PHYTOPHTHORA ROOT ROT AND BUCKEYE ROT

Cause

Several soil-borne *Phytophthora* species cause root rots of tomato in different parts of the world. In Australia, *Phytophthora nicotianae* and *Phytophthora erythroseptica* have been reported to cause root rot.

Symptoms

A wet dark brown rot develops in the vascular system within the taproot and crown area. This brown discolouration extends up or down from the point of infection. Brown lesions may also develop on the stem above or below the soil line, and these may girdle the stem or roots. Eventually the roots decay, causing the plants to wilt and die.

Phytophthora fruit disease is called buckeye rot. Buckeye rot almost always develops on fruit that are in contact with infested soil. For indeterminate trellis varieties, fruit closest to the ground are at greatest risk of infection if splashed by contaminated soil. Fruit develop a greyish-green to chocolate-brown firm rot with an indefinite, water-soaked margin and often with broad, zonate markings. The surface of the rot is generally smooth and the skin is intact. Although the rot progresses well into the flesh, affected fruit are firm initially and only soften at a late stage of infection.

Source of infection and spread

The pathogen is a soil-inhabiting water mould and requires water for spore production and fruit infection. The disease is generally confined to wet areas where rapid disease development can occur. The spores of the pathogen move through the soil during rain or irrigation until they contact roots or splash on to fruit. Warm, wet weather favours infection and disease development.

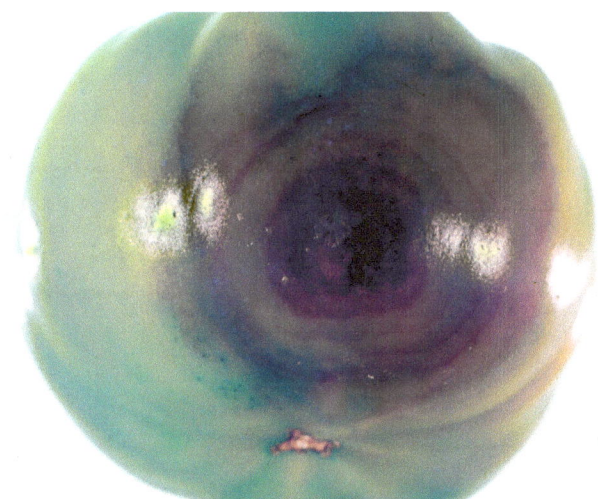

Fig 23.37 Buckeye rot on tomato fruit.

Importance

This is usually a minor disease, most likely to occur in wet areas.

Management

- Avoid planting tomatoes in infested areas, particularly low-lying areas of fields.

■ POWDERY MILDEW

Cause

Worldwide, three fungal species cause powdery mildew on tomato: *Leveillula taurica* (anamorph=*Oidiopsis* sp.), *Oidium lycopersici* and *Oidium neolycopersici*. All three species have been recorded in Australia.

Symptoms

Leveillula taurica initially causes light green, irregularly shaped leaf spots which progress to become yellow and then brown with age. Spots appear first on the older leaves of the plants and young leaves only become infected as they mature. Symptoms resemble those of leaf mould, but powdery mildew lacks the thick, velvety spore masses on the undersides of leaves that are a characteristic of leaf mould.

Fig 23.38 Early symptoms of powdery mildew (*Leveillula taurica*) on tomato leaves.

Fig 23.39 Advanced leaf infection by powdery mildew (*Leveillula taurica*).

Sporulation of *Leveillula* is confined to the underside of the leaves, but it often only seen with difficulty. Defoliation may follow severe outbreaks and yield and quality losses occur due to sunburn on the fruit.

Oidium lycopersici and O. neolycopersici cause grey-white powdery colonies to develop on both upper and lower leaf surfaces, as well as the stems and petioles. The underlying leaf tissues may initially turn purple, but later become yellow and then brown. Leaf twisting and deformation may result from severe infection.

Fig 23.40 Severe field infection by powdery mildew. Note the leaf chlorosis and necrosis.

Fig 23.41 Early leaf symptoms of powdery mildew caused by *Oidium lycopersici*.

Source of infection and spread

Powdery mildew fungi require a living host to survive. They persist on volunteer tomato plants, native crop and weed hosts and possibly as cleistothecia from their respective perfect stages. The fungi spread as air-borne spores. Spore germination is favoured by temperatures less than 30°C and intermediate (50–70%) to high (80–90%) relative humidity. Once an infection is established in a tomato leaf, temperatures greater than 30°C can accelerate both symptom development and the death of the leaf tissue. If environmental conditions favourable for powdery mildew infection persist throughout the growing season, inoculum pressure will gradually increase and hence the disease will tend to be most severe at the end of the cropping period.

Importance

Powdery mildew is very common and may cause severe losses during warm, dry weather. Severe outbreaks of powdery mildew caused by *Leveillula* can cause extensive

Fig 23.42 Stem and leaf symptoms of powdery mildew caused by *Oidium*.

leaf shrivelling, premature fruit ripening and severe sunburning in maturing fruit.

Powdery mildew caused by *Oidium* is more likely to be serious in glasshouse crops, where lower light intensity and high relative humidity favour disease development.

Management
- Spray with the recommended fungicides. Initiate a preventative spray schedule based on regular protectant fungicide applications in warm, dry weather.
- Remove old crops promptly.

■ RHIZOCTONIA FRUIT ROT

Cause
The fungus *Rhizoctonia solani*.

Symptoms
Small, circular, brown spots with definite concentric ring markings develop on green fruit. As the fruit ripen and the affected areas enlarge, the ring markings may disappear, with the spots becoming dark brown and the centre breaking open. A brownish mould often develops on the surface of the spots.

Source of infection and spread
The fungus is a common soil inhabitant attacking a wide range of plants. It affects only fruit on or close to the soil. Warm, wet weather favours infection and disease development.

Importance
This disease is not common, particularly in the current indeterminate trellised varieties where fruit is no longer in contact with the ground. When infection occurs, it can

result in extensive breakdown of affected fruit during transit and storage, particularly if fruit are not cool-stored.

Management
- Do not pack fruit showing signs of infection.
- Cool fruit promptly after harvest.

■ SCLEROTINIA ROT

Cause
The fungus *Sclerotinia sclerotiorum*.

Refer to the chapter on Common diseases of vegetable crops.

■ SCLEROTIUM BASE ROT

Cause
The fungus *Sclerotium rolfsii*.

Refer to the chapter on Common diseases of vegetable crops.

Fig 23.44 Sclerotium base rot in tomato. Note the dark, 'seed-like' sclerotes and fungal mycelium.

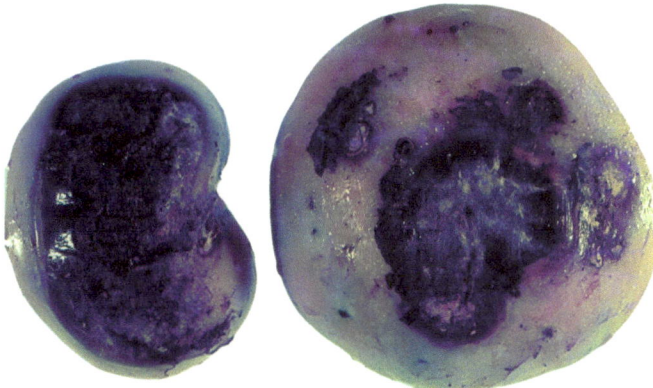

Fig 23.43 Rhizoctonia symptoms in tomato fruit.

Fig 23.45 Sclerotium infection in tomato fruit.

■ SEPTORIA LEAF SPOT

Cause

The fungus *Septoria lycopersici*.

Symptoms

The disease appears as small leaf spots, 3–4 mm wide, with brown margins and a light grey centre studded with small, black fruiting bodies (pycnidia) of the fungus. Severely affected leaves may yellow. Symptoms usually appear first on the older leaves.

Source of infection and spread

The fruiting bodies of the fungus contain large numbers of spores that spread in wind and water. Warm, wet weather favours infection and disease development. The fungus survives on undecomposed tomato residues in the soil.

Importance

Septoria leaf spot is not usually a problem in tomato crops in Australia.

Management

- Spray with the recommended fungicides.

- Disease forecasting models such as TOMCAST may be useful to help time fungicide applications.

■ TARGET SPOT (EARLY BLIGHT)

Cause

The fungus *Alternaria solani*.

Symptoms

Target spot occurs on the foliage, stem and fruit of tomato plants. The disease is first visible as small, brownish-black spots on older leaves. These develop into dark brown, ring spots with definite margins and yellow edges. The spots are commonly up to 6 mm wide, but may be as large as 12 mm when weather conditions are particularly favourable. When spotting is abundant, entire leaves may turn yellow and the plant may become defoliated, exposing the fruit to sunscald.

Stem spots resemble those on leaves but tend to be more elongated and the 'target' appearance is more pronounced. An affected seedling usually shows a large, dark, dry, sunken area on the stem near ground level. The plant is stunted and may break at this point.

Spots on fruit are black or dark brown, and are oval to round. These are common on the edge of the stem scar or around growth cracks. Spots develop rapidly as the fruit ripens.

Fig 23.46 Septoria leaf spot in tomato.

Fig 23.47 Target spot on tomato stems.

Fig 23.48 Symptoms of target spot on a tomato leaf.

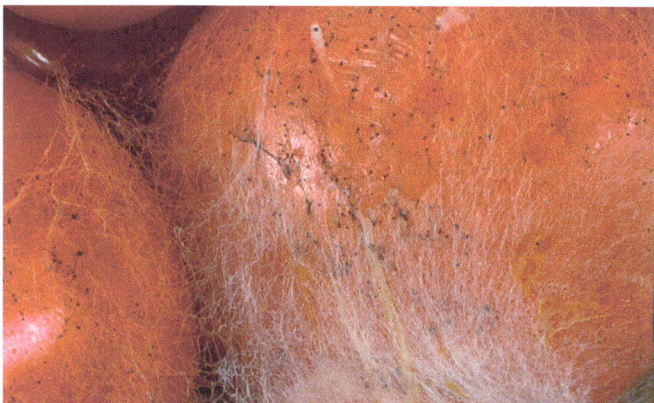

Fig 23.49 Transit rot (rhizopus soft rot) on tomato fruit.

Source of infection and spread

Large numbers of spores produced on leaf and stem spots spread from plant to plant in wind, rain and irrigation. Warm weather (temperature range 24–29°C) favours the disease, but it is still present to some extent throughout the year and may spread even in relatively dry weather. It develops more rapidly when plants are under stress. The fungus survives in infected crop residues, on other hosts such as potato, and on tomato seed.

Importance

Target spot is a common and serious disease requiring regular fungicide applications.

Management

- Use disease-free seed.

- Plant into well-prepared seedbeds or into soil-less mixes.

- Spray with the recommended fungicides in a preventative program prior to symptom development in both the seedling and field stages of growth.

- Disease forecasting models such as TOMCAST may be useful to help time fungicide applications.

- Rotate crops so that tomatoes do not follow either tomatoes or potatoes.

- Destroy old crops immediately after harvest.

■ TRANSIT ROT (RHIZOPUS SOFT ROT)

Cause

The fungus *Rhizopus stolonifer*.

Symptoms

Soft, slightly water-soaked spots with little discolouration develop rapidly over the entire fruit. The skin may remain intact with little sign of external mould, but affected fruit frequently split and collapse into a soft mass, that is quickly overtaken by the fungus. Black-stalked fruiting bodies develop in this growth.

Source of infection and spread

The fungus is common in soil, in the atmosphere and on decaying plant material. Fungal spores spread in air movement. Infection usually occurs through wounds, but in storage, the fungus may spread from fruit to fruit by contact. The disease is mainly a post-harvest problem and the fungus may become established in and around the packing shed. Heavy rains just before and during harvesting, hasten rot development because of increased fruit cracking, and mechanical damage from soil adhering to the fruit. High storage temperatures also favour disease development.

Importance

Transit rot is a common cause of post-harvest loss in tomatoes.

Management

- Handle fruit carefully to avoid skin damage.

- Discard fruit with serious growth cracks and other injuries.

- Remove reject fruit from the vicinity of the packing shed.

- Treat fruit with the recommended post-harvest fungicide.

■ VERTICILLIUM WILT

Cause

Worldwide, two species of *Verticillium* cause Verticillium wilt of tomatoes: *Verticillium dahliae* and *V. albo-atrum*. *Verticillium dahliae* is the more common pathogen and it is the only pathogen responsible for the disease in Australia.

Fig 23.50 Verticillium wilt in field tomato plants.

Symptoms

The first sign of Verticillium wilt is often a wilting of the oldest leaves during the warmest part of the day, with recovery at night. The lower leaves turn pale or are blotchy with pale green or yellow areas. As the disease progresses, leaf margins and tips yellow, and chlorotic, angular, interveinal, V-shaped lesions develop. In time, the leaf margins and yellow areas on leaves become necrotic. Infected plants are stunted and unthrifty. If the lower stems are cut longitudinally, a brown discolouration of the water-conducting tissues is evident. The discolouration may also extend into the petioles of the lower leaves. Disease

Fig 23.51 Verticillium wilt in tomato. Note yellowing and necrosis in the lower leaflets.

symptoms may be accentuated if the plant is under stress due to fruit load or some other factor.

Source of infection and spread

The fungus is a common soil inhabitant and survives in the soil for long periods as microsclerotia. It enters the roots, usually through wounds, and moves into the water-conducting tissues of the stem. The disease is favoured by cool weather (optimal temperature range 20–24°C). Verticillium infects other crop plants and weeds including potato, cotton and peanut. Susceptible weeds include Noogoora burr (*Xanthium pungens*), stinking roger (*Tagetes minuta*) and cobbler's pegs (*Bidens pilosa*).

Importance

Two races of the fungus occur in Queensland. Many commercial varieties are resistant to Race 1, but resistance has not yet been identified for race 2. Race 2 is common and often severe in southern Queensland. Verticillium wilt is rarely seen in northern production areas.

Management

- Avoid growing potatoes and tomatoes in succession.

- Keep crop areas free from tomato volunteers, Noogoora burr and other susceptible weeds.

- Choose varieties with resistance to race 1.

- Take steps to avoid moving the fungus from infested to clean areas on vehicles and machinery.

■ YEASTY ROT

Cause

The fungus *Geotrichum candidum*.

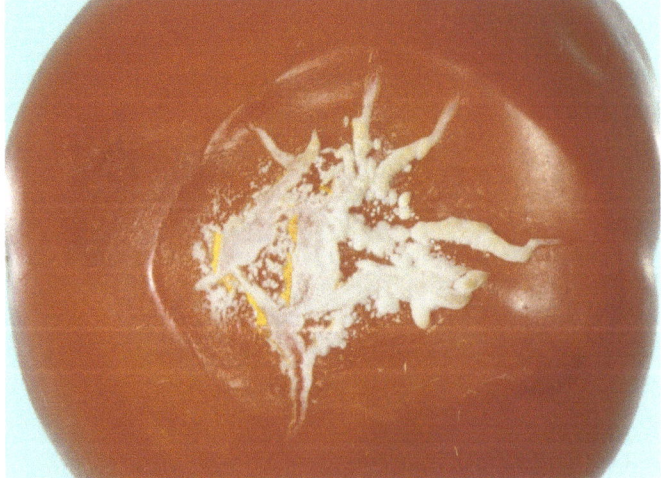

Fig 23.52 Yeasty rot in tomato.

Symptoms

On fruit, water-soaked to bleached areas develop, often beginning at cracks or injuries to the skin or at the stem scar. In green fruit, affected areas generally remain quite firm until the decay is well advanced. In ripening fruit, decay progresses rapidly, eventually causing complete collapse. Cracks in the skin over affected areas generally fill with a whitish, scum-like mass of fungus. Yeasty rot often occurs in association with bacterial soft rot.

Source of infection and spread

The fungus is common on decaying plant matter in the soil. It spreads to fruit in wind and water and on insects. The fungus is a wound pathogen that cannot penetrate the fruit epidermis directly, although it causes a rapid fruit breakdown if it enters the fruit through wounds. Infection occurs through the stem scar or skin injuries. Hot, wet weather favours infection and disease development. Although most infection originates in the field, extensive spread may occur during harvesting and packing, particularly if there has been heavy rain, which increases fruit cracking and skin injuries from adhering soil. In packing sheds and processing plants, the fungus may grow on plant residues remaining on equipment. High storage temperatures favour rapid rot development. Good hygiene is important in the packing shed to prevent the fungus spreading in post-harvest dips.

Importance

Yeasty rot is a common disease during hot, humid weather.

Management

- Handle fruit carefully to avoid skin damage.

- Discard fruit with serious growth cracks and other injuries.

Fig 23.53 Severe root galling in tomato caused by root-knot nematode.

- Remove reject fruit from the vicinity of the packing shed.

- Treat fruit with the recommended post-harvest fungicides.

NEMATODES

■ ROOT-KNOT NEMATODE

Refer to the chapter on Common diseases of vegetable crops.

VIRUSES AND VIROIDS

BUNCHY TOP – BIOSECURITY THREAT

Cause

Potato spindle tuber viroid.

Symptoms

Mild and severe strains occur and mild strains may cause no obvious symptoms. With severe strains, the plant can be stunted and have a bunched appearance due to shortening of the internodes and shoot growth can be spindly. Leaf symptoms include yellowing and purpling, down curling and twisting. Younger leaves are relatively small, while middle and older leaves show vein necrosis and eventually die. Flowers may abort. Fruit size is reduced and fruit may ripen erratically and become hard and dark green.

Source of infection and spread

Long distance spread is usually through infected true seed of tomato and infected seed tubers of potato. The viroid reaches high concentrations in infected plants, and once established, spreads locally with mechanical contact such as cutting and pruning procedures, handling plants, plant-to-plant contact, and contaminated machinery, trellises, hands and clothing.

Fig 23.54 Field symptoms of bunchy top (*Potato spindle tuber viroid*).

Fig 23.55 Bunchy top (*Potato spindle tuber viroid*) in tomato.

Importance

The viroid occurs in many tomato and potato growing areas in the world and causes serious disease in both these crops. Sporadic outbreaks and eradication programs have occurred in several Australian States.

What to do if you suspect bunchy top

This pathogen is a biosecurity risk to Australia. Any suspected affected plants should be reported to the nearest Department of Primary Industries or the Plant Health Australia hotline (1800 084 881).

■ CAPSICUM CHLOROSIS VIRUS

Cause

Capsicum chlorosis virus (CaCV; Tospovirus).

Symptoms

Infected plants can be stunted and show large, irregular chlorotic and necrotic blotches on the leaflets. Leavers may be twisted and curled. Fruit on infected plants have chlorotic or necrotic patches and rings, and are often small and distorted.

Importance

First identified in 1999 the virus is now prevalent in tomato and capsicum crops in coastal Queensland.

For further information, refer to the chapters on Capsicum diseases and Common diseases of vegetable crops.

Fig 23.56 *Capsicum chlorosis virus* in tomato leaves.

Fig 23.57 *Capsicum chlorosis virus* in tomato fruit.

Fig 23.58 *Capsicum chlorosis virus* showing various symptoms in tomato fruit.

Fig 23.59 Fern leaf (*Cucumber mosaic virus*) in tomato.

■ FERN LEAF

Cause
Cucumber mosaic virus (CMV; Cucumovirus).

Symptoms
The first symptom is a thickening and rolling of the leaf edges. Later, the terminal shoots become a mass of very narrow, distorted leaflets, all with thickened and curled edges. Fruit on affected plants may be malformed.

Source of infection and spread
The virus has a wide range of hosts among crop and weed species. It is spread from plant-to-plant by aphids in a non-persistent manner.

Importance
Fern leaf is a minor disease, but regularly occurs in some districts.

Management
Management is not warranted.

■ LEAF SHRIVEL

Cause
Potato virus Y (PVY; Potyvirus).

Symptoms
Young leaflets and petioles curl downwards, which gives the leaf a 'clawed' appearance. Leaves may also be mottled,

Fig 23.60 Leaf shrivel (*Potato virus Y*) in tomato.

and the leaf area of the plant reduced. Older leaves may show a dark grey to brown spotting on the underside, and eventually shrivel and die. Fruit show no symptoms, although total yield is reduced and fruit may be damaged by sunburn due to reduced leaf cover.

Source of infection and spread

The virus is transmitted by several species of aphid in a non-persistent manner. The main sources of infection are old infected tomato, capsicum and tobacco crops, and weeds such as gooseberry (*Physalis* spp.), nightshade (*Solanum* spp.) and apple of Peru (*Nicandra physalodes*).

Importance

Leaf shrivel is widespread and often severe, especially late in the season in Queensland.

Management

- Destroy old tomato crops and related species.

- If capsicums are grown concurrently, use a capsicum cultivar resistant to the virus.

- Site the seedling production areas well away from older tomato crops.

- Control aphids with the recommended insecticides.

PEPINO MOSAIC – BIOSECURITY THREAT

Cause

Pepino mosaic virus (PepMV; Potexvirus).

Symptoms

Symptoms are variable and depend on the variety, weather conditions and light intensity. Symptoms are likely to be more intense during cool, cloudy weather. Infected plants have distorted leaf development, yellow leaf spots, mosaic and bubbling of the leaf surface. Plants may develop a lighter green, spiky or nettle-like head with reduced leaf size at the apex of the plant. Virus infection results in reduced fruit size and quality. Fruit may ripen unevenly and develop blotches and marbling on the skin.

Source of infection and spread

The virus is highly contagious and is readily spread by contact, contaminated tools, hands or clothing, grafting and plant-to-plant contact. The virus can remain viable and infectious on contaminated equipment and clothing for several weeks, and can survive in crop debris for at least several months. *Pepino mosaic virus* can be spread by both hand pollination and by bumblebees used for pollination in glasshouse-grown crops. The virus is also spread with seed coat contamination, resulting in a low rate of seed transmission, which is sufficient to introduce the virus into a new location or nursery with subsequent spread by contact. There is no evidence that common virus vectors such as aphids and whiteflies can transmit *Pepino mosaic virus*.

Importance

Pepino mosaic virus is of increasing concern in glasshouse tomato production in Europe, North and South America. The virus, once established, is difficult to manage with sanitation and roguing being the main management tools.

What to do if you suspect Pepino mosaic virus

This pathogen is a biosecurity risk to Australia. Any suspected affected plants should be reported to the nearest Department of Primary Industries or the Plant Health Australia hotline (1800 084 881).

Fig 23.61 Leaf symptoms of *Pepino mosaic virus*.

Fig 23.62 Fruit symptoms of *Pepino mosaic virus*.

■ SPOTTED WILT

Cause
Tomato spotted wilt virus (TSWV; Tospovirus).

Symptoms
Young leaves usually turn a bronze colour and develop numerous small, dark spots. Leafstalks may bend downwards and plants may have a one-sided growth habit. Alternatively, they are entirely stunted with drooping leaves. Affected leaves may wither and die. Dark streaks often appear on stems near the growing point. Ring spots, which may be white to brown in colour, develop on ripening fruit. Fruit is often cracked, distorted and reduced in size.

Importance
TSWV can cause important crop losses in most Australian tomato production areas.

For further information, refer to the chapter on Common diseases of vegetable crops.

Fig 23.64 Plant symptoms of *Tomato spotted wilt virus*.

Fig 23.63 Leaf symptoms of *Tomato spotted wilt virus* (TSWV).

Fig 23.65 *Tomato spotted wilt virus* on green fruit.

Fig 23.66 *Tomato spotted wilt virus* on Roma type tomatoes.

■ TOMATO MOSAIC

Cause
Tomato mosaic virus (TMV; Tobamovirus).

Symptoms
Affected plants are generally lighter in colour than healthy plants. Individual leaves are slightly crinkled, showing a distinct light green and dark green mosaic, particularly at low temperatures. The virus also causes a pale blotching and internal browning of the fruit.

Source of infection and spread
Tomato mosaic virus is persistent and highly infectious. The main sources of infection are contaminated seed, other affected tomato crops or undecomposed tomato residues in the soil, old trellis material and contaminated hands and

Fig 23.68 Internal, brown discolouration of tomato fruit caused by *Tomato mosaic virus*.

clothing. Crops and weeds related to tomato, such as tobacco, Cape gooseberry and blackberry nightshade, may also be sources of the virus.

The virus spreads readily on hands, implements and pruning knives and by direct contact between plants.

Importance
TMV may be serious but controlled effectively if all precautions are followed.

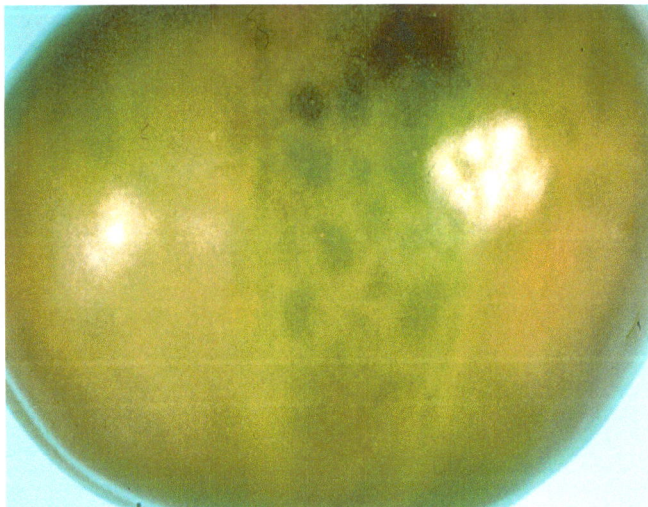

Fig 23.67 *Tomato mosaic virus* on fruit.

Fig 23.69 *Tomato mosaic virus* symptoms on tomato leaves.

Management

- Destroy residues of old tomato crops and other susceptible crops as soon as possible.

- Use commercially treated seed or treat seed with 10% trisodium phosphate (TSP) for one hour.

- Avoid handling seedlings after working in an older crop or, if this is not practicable, wash hands thoroughly and dip in 10% TSP.

- Treat pruning knives and implements with 10% TSP after working in a diseased crop. Wires and posts should also be treated before re-use.

- Use varieties carrying tomato mosaic resistance genes.

◼ TOMATO YELLOW LEAF CURL/TOMATO LEAF CURL VIRUS

Cause

Tomato yellow leaf curl virus (TYLCV; Begomovirus).

Tomato leaf curl virus (ToLCV; Begomovirus).

Symptoms

Both viruses produce similar symptoms on tomato plants.

Affected plants grow slowly and become stunted or dwarfed. Leaflets are rolled upwards and inwards and display interveinal chlorosis and rugosity of the leaf blade, and can be rounded. Leaves are often bent downwards and are stiff. The flowers may wither. Fruit, if produced at all, are small, dry and unsaleable.

Source of infection and spread

TYLCV is spread by the silverleaf whitefly (*Bemisia tabaci* biotype B). Although the immature nymphal stages of the

Fig 23.70 *Tomato yellow leaf curl virus* showing symptoms on new growth.

Fig 23.71 *Tomato yellow leaf curl virus* showing erect leaves.

whitefly can acquire virus from infected plants, it is the active, adult insects that are responsible for almost all virus spread into and within crops. The whitefly acquires the virus while feeding by sucking plant sap through a stylet. The insects need to feed on infected plants for at least 15 minutes to acquire virus and then feed for 15–30 minutes to transmit the virus to another host plant. Transmission efficiency increases as the duration of the feeding times increase.

The tomato leaf curl viruses are not transmitted by aphids, leafhoppers or leaf-eating pests such as beetles or grasshoppers.

Other hosts of TYLCV include French beans and thornapple (*Datura stramonium*) in which strong leaf curling and interveinal chlorosis is produced, and capsicum, which is infected without symptoms. Cotton, eggplant and poinsettia are not infected.

Old infected crops are a major source of infection, but, at present, the major weed or crop hosts that allow the virus to survive between tomato growing seasons are unknown.

Tomato leaf curl virus can be transmitted by the silverleaf whitefly. It is also transmitted by the Australian native biotype of *Bemisia tabaci* that feeds, but does not reproduce on tomato.

In addition to tomato, tomato leaf curl can infect *Nicandra physalodes*, *Datura stramonium* (common thornapple), *Physalis virginiana* and several *Nicotiana* and *Solanum* species.

Importance

Tomato yellow leaf curl is one of the most important tomato diseases worldwide. At present, it only occurs in south-eastern Queensland, but has the potential to spread to all tomato-growing areas that provide an environment suited to the silverleaf whitefly.

Tomato leaf curl virus-Australia has been present in coastal northern Australia for at least 40 years, causing sporadic, serious losses in tomato crops in the Northern Territory. In Queensland, the virus has been found on several occasions on Cape York Peninsula but not in the major tomato production areas. The endemic whitefly vector of this virus has a narrow host range and a limited capacity to disperse.

Management

Tomato leaf curl diseases are difficult to control. Eliminating susceptible weed species, isolating new crops from older infected crops and controlling the whitefly vectors can reduce disease.

Several sources of resistance are available in tomato to TYLCV, and varieties carrying the *Ty-1* gene have good resistance to TYLCV in Australia.

■ TOMATO YELLOW TOP

Cause

Potato leafroll virus (syn. *Tomato yellow top virus*) (Polerovirus).

Symptoms

Infected plants are stunted, assume a stiff, upright appearance and appear markedly chlorotic from a distance. Symptoms include reduced leaf size, rounding and marginal chlorosis of leaflets, and down curling of leaflet margins. Flowers often fail to set fruit and flower buds can be killed following infection. Yield can be reduced drastically if plants are infected at an early stage, but late infection has progressively less effect. The fruit do not produce symptoms.

The symptoms are similar to those *of Tomato yellow leaf curl virus.*

Source of infection and spread

The most important insect vector is the green peach aphid (*Myzus persicae*), although the potato aphid (*Macrosiphum euphorbiae*) is often involved. Other hosts of the virus include potato and the weeds thornapple (*Datura* spp.), apple of Peru (*Nicandra physalodes*), nightshade (*Solanum* spp.) and shepherd's-purse (*Capsella bursa-pastoris*).

Importance

Yellow top occurs occasionally, especially during autumn, in tomato crops in coastal areas of Queensland.

Management

The disease is difficult to control. Eliminating susceptible weed species, isolating new crops from older infected crops and controlling the aphid vectors can reduce disease levels.

■ TOMATO TORRADO DISEASE

Cause

Tomato torrado virus (Picorna-like virus).

Symptoms

The initial symptoms are necrotic spots, surrounded by a light green or yellow area at the base of the leaflets. These affected areas may fall out, leaving holes in the leaflets. Necrosis and mottling extend to the remainder of the leaves. In highly susceptible varieties, the virus causes severe necrosis and death of leaves, named 'torrado' by Spanish growers, meaning burned or roasted.

Fig 23.72 *Tomato yellow top virus* showing leaf damage.

Fig 23.73 Leaf symptoms of *Tomato torrado virus.*

Fig 23.74 *Tomato torrado virus* showing basal discolouration in leaves.

In Spain, the virus may also cause necrotic rings and seams on fruit, making them unmarketable.

Source of infection and spread

The virus is spread by both the glasshouse whitefly (*Trialeurodes vaporariorum*) and the silverleaf whitefly (*Bemisia tabaci*).

Capsicum and eggplant are also hosts in Europe and the virus infects a range of common broadleafed weed species.

Importance

Tomato torrado disease, first found in Spain in 2004, is a serious disease of field and glasshouse grown tomatoes in Spain, the Canary Islands and Mexico where a virus closely related to Tomato torrado virus occurs.

The virus was identified from glasshouse-grown tomatoes on the North Adelaide Plain of South Australia in 2008 and is known to have been present since 2006.

In Australia, the virus causes a reduction in fruit set and fruit size in infected plants. Disease incidence varies with the time of planting and variety.

Fig 23.75 *Tomato torrado virus* on a glasshouse plant.

Management

- Control weeds both in and around crops as these may harbour both the virus and whitefly vectors.

- Destroy crops promptly after harvest.

- Plant varieties that are resistant or tolerant to the virus, in situations where it is likely to be prevalent.

■ FURTHER INFORMATION

Jones JB, Jones JP, Stall RE and Zitter TA (Eds) (1991). *Compendium of tomato diseases*. APS Press, St Paul, Minnesota.

GLOSSARY

acervulus: saucer-shaped fungal fruiting body that produces conidia; a characteristic of the genus *Colletotrichum*. Plural – acervuli.

anamorph: the asexual or imperfect state in the life cycle of a fungus in which asexual spores, such as conidia, are produced.

anastomosis: the union of fungal hyphae to form a network that allows genetic material to be shared.

anastomosis group: isolates or strains with hyphae that are able to fuse and share cellular contents with one another.

anthracnose: a plant disease having characteristic depressed spots or lesions on leaves, stems or fruit. Usually caused by fungi that produce asexual spores in an acervulus.

apothecium: an open-cup or saucer-shaped fungal structure producing ascospores.

appressorium: a structure produced by a germinating fungal spore that adheres to the surface of the host and helps the fungus penetrate the surface cells.

ascospore: a spore formed during the sexual stage in the life cycle of certain fungi and carried in a microscopic structure called an ascus.

asexual reproduction: reproduction not involving the sex organs or sex cells; a common form of reproduction in fungi and bacteria.

bacterium: a single-celled microorganism lacking chlorophyll and capable of extremely rapid reproduction.

bark: the plant tissue outside the cambium in vascular plants.

biovar: a group of genetically identical individuals within a species.

blight: a disease typified by general and rapid death of leaves, flowers or stems.

blotch: a disease typified by large, irregular spots or blots on plant parts.

calyx: the outer group of leaves surrounding a flower, often small and green-coloured.

cambium: a thin layer of actively dividing cells located between the xylem (water-conducting) and phloem (food-conducting) systems in most vascular plants. The cambium is responsible for the growth in the diameter of plants.

canker: a dead, often sunken or cracked, area on a stem, twig, limb or trunk and surrounded by healthy tissue.

chlamydospore: a thick-walled resting spore of a fungus which allows survival during adverse conditions.

chlorophyll: an organic compound giving the green colour to plants and used to make food during photosynthesis.

chlorosis: a partial or complete absence of normal green colour from plant parts.

coalesce: to run together.

conidium: a fungal spore produced during the non-sexual stage of the life cycle.

corm: a short, swollen underground stem.

crown: a shortened stem growing close to the ground with leaves and axillary buds.

cyst: a tough, leathery structure containing a dead female nematode and eggs. Also formed by some fungi and producing motile spores.

cytoplasm: all of the cell's living substances except the nucleus and cell wall.

damping-off: the rapid death of germinating seed or seedlings before or after emergence.

detasselling: the removal of male flowers of sweet corn to prevent self-pollination or inbreeding.

dieback: the progressive death of roots, shoots or branches generally starting at the tip.

epidemic: a widespread and severe outbreak of an infectious disease.

epiphyte: any organism living on plant surfaces without causing infection.

forma specialis: abbreviation f.sp. A taxonomic group within a pathogen species defined in terms of host range, i.e. members of different formae speciales infect different groups of plants.

fruiting body: a general term for spore-bearing structures of fungi.

fumigant: a toxic gas or volatile substance used to control pests in a certain area.

fungicide: a chemical compound that kills or inhibits fungi. A protectant fungicide provides a protective chemical barrier over the host surface and prevents initial infection. An eradicant fungicide kills fungal pathogens growing within the host and so has a curative effect. Eradicant fungicides are often absorbed and transported internally within the plant.

fungus: a type of organism lacking chlorophyll and having individual strands (hyphae) aggregated into mycelium. Some cause plant diseases.

gall: a swelling or outgrowth produced on a plant as a result of attack by pathogens or insects.

host plant: a plant that is invaded by a pathogen and from which the pathogen obtains nutrients.

host range: the various kinds of plants attacked by a certain pathogen.

hydathode: a small, natural opening or pore that discharges water from the interior of a leaf to the surface; generally occurring around leaf margin and larger than a stoma.

hypha: one of the threads of a fungal mycelium.

imperfect state: see anamorph.

infection: the establishment of a pathogen within a host plant.

inoculum: pathogen or pathogen parts capable of infecting a host plant. Examples include fungal spores, bacterial cells and virus particles.

instar: an insect stage between moults before adulthood.

larva (pl. larvae): the juvenile stage of certain animals e.g. insects, nematodes.

latent infection: infection of a host by a pathogen without symptom development.

lenticel: a minute hole or pore in some plant parts allowing gas and water vapour exchange.

lesion: a localised area of diseased tissue.

life cycle: the successive stages in the growth and development of an organism that occur between the appearance and reappearance of the same stage.

microsclerotia: very small sclerotia.

mosaic: patchy variation of normal green colour of leaves; a common symptom of virus infection in plants.

mottle: an irregular pattern of light and dark green areas.

mummify: to dry and shrivel. A mummy is a dried and shrivelled fruit.

mycelium: the mass of individual threads or hyphae that make up the body of a fungus.

necrosis (adj. necrotic): death of plant cells and tissues, usually accompanied by discolouration of the affected area.

nematicide: a chemical compound that kills or inhibits nematodes.

nematode: a small, worm-like animal that is parasitic on plants and other animals, or free-living in soil or water.

oomycete: a group of plant pathogens once considered fungi but now placed in the Kingdom Straminipila. The name derives from the microscopic round oospores produced in sexual reproduction. Members are well adapted to water and moist conditions, and most produce motile zoospores with flagellae that enable them to swim. Important plant pathogens in the group include *Phytophthora, Pythium* and the downy mildews.

oospore: a thick-walled resting spore formed during sexual reproduction in some fungi.

parasite: an organism living in or on an other living organism and obtaining its food from the latter.

pathogen: a parasite able to cause disease in a host. The major plant pathogens are fungi, bacteria, nematodes and viruses.

pathogenicity: the ability of a pathogen to cause disease.

pathotype: a subdivision of a species distinguished by the ability to infect certain host species or cultivars. Generally used in relation to fungal pathogens, whereas pathovar usually refers to bacterial pathogens.

pathovar (pv.): in bacteria, a sub-species or group of strains that can infect only plants within a certain genus or species.

perfect stage: see teleomorph.

pericarp: the wall of a fruit.

perithecium: a microscopic, globular or flask-shaped structure containing sexual fungal spores. Plural - perithecia.

petiole: the stalk of a leaf.

pH: symbol of a scale (0–14) used to indicate acidity or alkalinity; 0–6 indicates acidity, 7 is neutral and 8–14 indicates alkalinity.

phloem: the food-conducting tissue of a plant.

phytoplasma: an obligate single-celled organism that lacks a cell wall. Formerly: mycoplasma.

primary inoculum: an inoculum, usually from an overwintering source, that initiates disease each season in contrast to the secondary inoculum that spreads the disease during the season.

PT material: pathogen-tested material.

pustule: a blister-like, raised spot on plant tissue, erupting to release fungal spores, e.g. rust diseases.

pycnidium: a microscopic, flask-shaped fruiting body producing asexual spores. Plural – pycnidia.

race: a genetically and often geographically distinct group within a species. Also a group of pathogens that infect a given set of plant varieties.

ratoon: cutting back mature plants to rejuvenate them and obtain a second crop from the same plants.

resistance (adj. resistant): the ability of a host to prevent or reduce the development of a pathogen.

rhizome: an underground stem which grows horizontally and may be used for storage or vegetative propagation.

rogue: to remove a diseased plant or plants with the aim of reducing disease spread.

rosette: short, bunchy habit of plant growth.

russet: brownish, roughened areas on the skin of fruit.

saprophyte: an organism that uses dead organic material for food.

scab: crust-like, diseased area on the surface of a plant part.

sclerotium: a small, hard, dark fungal structure that can survive for long periods under unfavourable environmental conditions: an important means of survival for many pathogens, e.g. *Sclerotium rolfsii*.

secondary inoculum: inoculum produced by infections that occurred during the same growing season.

senesce (noun, senescence)**:** to decline with maturity or age.

shot hole: a symptom of disease in which small circular leaf spots fall out and leave holes in their place.

soil inhabitants: microorganisms that are able to survive indefinitely in soil as saprophytes.

species: a class of individuals usually able to interbreed freely and having many characteristics in common. The abbreviations 'sp.' and 'spp.' used after a genus name refers to (respectively) one or several undetermined species without naming them individually.

sporangium: a fungal structure containing spores produced during the asexual stage of the life cycle.

spore: the reproductive unit of a fungus which functions as a seed.

sporidia: the basidiospores of the smut fungi.

sporulation: the production of spores.

stipule: a small, leaf-like structure at the base of a leaf or along a leafstalk.

stoma: a breathing pore in leaves, stems, and some fruit allowing the exchange of gases between the plant and the atmosphere.

suture: the junction of two contiguous parts, e.g. the junction of the surfaces of a pea pod.

symptom: the indication of a disease by the reaction of the host.

synnemata: a compact group of erect conidiophores bearing conidia at the apex only or at the apex and on the sides.

teleomorph: the sexual form or perfect state in the life cycle of a fungus, in which sexual spores (e.g. ascospores) are formed.

teliospore: a sexual, thick-walled resting spore of rust or smut fungi.

tissue: a group of cells, usually of similar structure, performing the same or related functions.

tuber: a short, thickened, fleshy underground stem, typically produced at the end of a stolon, e.g. potato tuber.

uredinospore: a rust spore produced in an uredium.

vascular: associated with the conductive tissue (xylem and phloem) of plants.

vector: a living organism that transmits a pathogen.

vegetative propagation: the use of cuttings, bulbs, tubers, corms and other vegetative plant parts to grow new plants.

veinclearing: an increased translucency of the veinal system in a leaf, making the pattern more pronounced, light against dark, by transmitted light.

virus: an extremely small pathogen consisting of infectious nucleic acid and a protective protein coat. A virus can multiply only within the living cells of a host.

ware potatoes: potato tubers harvested for human consumption in contrast to 'seed' potatoes that are planted to propagate the crop vegetatively.

water-conducting tissue: xylem vascular tissue that transports water and minerals from the root system to other parts of a plant.

water-soaked: diseased areas that appear congested with water.

wilt: a loss of rigidity and drooping of plant parts due to inadequate water supply or excessive water loss by a plant.

xylem: water-conducting and supporting tissue of roots, stems and other plant parts.

zonate: containing a concentric ring pattern of variation in spore production, texture or pigmentation.

zoospore: a spore with flagella or tails, and capable of moving in water.

INDEX